Conservation of Germplasm Resources

AN IMPERATIVE

Committee on Germplasm Resources
Division of Biological Sciences
Assembly of Life Sciences
National Research Council

NATIONAL ACADEMY OF SCIENCES
Washington, D.C.   1978

This study was supported by Contract No. EY-76-C-02-2708-007 with the Energy Research and Development Administration (now Department of Energy).

International Standard Book Number 0-309-02744-6

Library of Congress Card Catalog Number 78-54007

Available from
Printing and Publishing Office
National Academy of Sciences
2101 Constitution Avenue
Washington, D.C. 20418

Printed in the United States of America

PREFACE

In September 1976 the Assembly of Life Sciences, National
Research Council, established a Committee on Germplasm
Resources and assigned it these tasks:

- To study the status of germplasm resources
- To assess efforts to solve the problems associated
with conservation of germplasm resources
- To prepare a report containing recommendations for
making these efforts more effective.

The Committee has prepared its report for consideration
by government officials, scientists (including those whose
skills and experience are in areas not immediately related
to the central issue of germplasm), and concerned citizens
generally. The report is not intended to be encyclopedic,
but underscores the fact that genetic diversity for many
species is severely threatened and that, although considerable effort is already devoted to preservation of germplasm, much more emphasis is needed.

Soon after commencing its deliberations, the Committee
sought help from a number of specialists in assessing the
status of various types of organisms. Many responded
graciously to these requests for information. We wish to
express special appreciation to the following persons:
Robert Bye, University of Colorado; Richard Donovick,
Director, American Type Culture Collection; Norman R.
Farnsworth, University of Illinois Medical Center; Tom
Gilbert, National Park Service, U.S. Department of the
Interior; Howard S. Irwin, President, New York Botanical
Garden; Leon Jacobs, National Institutes of Health, U.S.
Department of Health, Education, and Welfare; Robert
Jenkins, Vice President for Science, The Nature Conservancy;
H. E. Kennedy, Biological Abstracts; Stanley L. Krugman,

Forest Service, U.S. Department of Agriculture; Arnold L.
Lum, Woods Hole Oceanographic Institute; Nancy A.
Muckenhirn, Institute of Laboratory Animal Resources,
National Research Council; Richard L. Saunders, North
American Salmon Research Center, New Brunswick, Canada;
William E. Sievers, National Science Foundation; Claire E.
Terrill, Agricultural Research Service, U.S. Department
of Agriculture; and H. Garrison Wilkes, University of
Massachusetts.

Committee on Germplasm Resources

COMMITTEE ON GERMPLASM RESOURCES

Elizabeth S. Russell, The Jackson Laboratory, Bar Harbor, Maine (*Chairman*)

Barbara Bachmann, Department of Human Genetics, Yale University School of Medicine, New Haven, Connecticut

Kurt Benirschke, Department of Pathology, University of California, San Diego

Robert W. Briggs, Department of Zoology, Indiana University, Bloomington

Richard H. Goodwin, Connecticut College, New London

Judith Grassle, Marine Biological Laboratory, Woods Hole, Massachusetts

Burke H. Judd, Department of Zoology, University of Texas, Austin

Charles F. Lewis, Agricultural Research Service, U.S. Department of Agriculture, Beltsville, Maryland

Peter Mazur, Biology Division, Oak Ridge National Laboratory, Oak Ridge, Tennessee

David L. Nanney, Department of Genetics and Development, University of Illinois, Urbana

Thomas H. Roderick, The Jackson Laboratory, Bar Harbor, Maine

David J. Rogers, Department of E.P.O. Biology, University of Colorado, Boulder

Staff Officer: Veronica I. Pye

CONTENTS

1 INTRODUCTION

Germplasm resources may be defined as the total array of living species, subspecies, genetically defined stocks, genetic variants, and mutants whose continuing availability is important for society's present and future health and welfare. They may be regarded as the biological under-pinning on which we live.

The earth is populated with millions of species of organisms that interact with one another to form complex ecosystems. Almost certainly there is considerable diversity within the gene pool of each of these species. Natural ecosystems and their component species have been evolving over millions of years without human influence and until comparatively recently have been unaffected by that influence. Not all the species in many natural ecosystems have been identified, nor do we understand many of the interactions in these systems; we find it difficult to assign them a precise quantitative value, anticipate the impact of human activities, modify that impact, and determine to what extent such perturbations are inevitable. Until recently the earth's natural habitats were so extensive that most people felt the preservation of whatever component species were of practical use to humans could be taken for granted.

With increasing industrial expansion and the rapid growth of human populations, incurring greatly increased needs for food, living space, and sources of energy, the world is losing many important natural habitat areas. We may be in great danger of losing species and variants that are essential to meet long-term human needs. There are likely to be, among these, species whose value we already know and others whose importance has not yet been recognized. Enlightened self-interest directs that we act immediately to prevent wholesale loss of resources that have evolved through millions of years of mutation and selection.

1

Environmental fluctuations and minor differences among
similar ecosystems have led to the selection and maintenance of many gene differences among scattered populations
of a given species of organism in nature. These differences
constitute a rich source of variability that can be ampli-
fied through recombination and selection. The sum of these
gene differences within a species constitutes the *gene pool*
of that species. It is this natural variability that man
has exploited in the past in developing domestic plants
and animals from wild populations through selective breeding. It is extremely important that this naturally occur-
ring richness of diversity within species be maintained.
For this reason, efforts at preserving germplasm must look
to the preservation of large populations of organisms,
rather than merely a few breeding individuals.

Over a very long period, man has affected the genetic
development of a significant, although relatively small,
number of species through the selection of agricultural
crops and domestic animals. We know that the gene pools
of the "folk varieties" lying behind what are now major
cereal crops contain many valuable genes that can be incorporated into modern genetically controlled stocks for
such purposes as increasing resistance to specific pathogens and ensuring hardiness under difficult environmental
conditions. Some of these genes from folk varieties,
or land races, which have important positive effects in
all environmental conditions, have already been selected
for the development of genetically homogeneous high-
production strains. Other genes, with effects needed only
under special circumstances or in environments not encountered during the initial selection process, may not
be incorporated into modern varieties. These "miracle"
crops, which provide impressive yields under favorable
conditions, have markedly increased cereal production in
many parts of the world, thus helping to avert crises in
the world food supply. This great accomplishment has its
negative aspects, however: In some areas there is a clear
danger that plantings of the new crops will completely
replace plantings of indigenous folk varieties. Should
this happen, important reservoirs of genes needed under
less-favorable conditions will be irretrievably lost unless
adequate provisions are made to guarantee their preservation.

Collections of varieites of economically important
organisms serve as reservoirs of genes. Highly selected
stocks of many economically important species have been

produced and characterized. Their value resides both in the beneficial effects of the genes they carry and in the basic knowledge that results from the considerable human effort involved in producing and characterizing them. Many of these valuable stocks will be lost unless responsibility for maintaining them is assigned to competent curators and funds are provided to support the work.

Advances in biological research frequently depend on the availability of pertinent research material. In some areas of research, current choices of research organisms require species found only in the wild, and decreased availability of these species becomes an impediment to research. Research in many fields depends on supplies of organisms from special genetic collections. Certain convenient, well-characterized species are widely used in basic genetic research. Critical experiments using these species frequently require specific genetically defined stocks carrying known arrays of mutations or chromosome rearrangements. Specific enzyme deficiencies or alterations are required in some biochemical research. A considerable array of mutant experimental animals with counterparts of certain human diseases serve as valuable tools in biomedical research, and new models are still needed. One recurrent problem is that of providing for maintenance of genetic collections and mutant research colonies beyond the periods of the careers of the persons who established them.

Whatever the example chosen, it is clear that in each case the main issue has to do with preservation of the basic genetic material, DNA, which has the unique quality of carrying information, and the capacity of these sequences to replicate themselves exactly. As long as we have one copy of a particular gene, we have in principle the capacity to make more. Once *all* copies of a gene have been lost, resynthesizing its identical DNA sequence and integrating it with all the other pertinent genes to constitute the formula for a viable and useful organism is well nigh impossible. Co-adapted gene sequences, which are at least as important as single genes in this context, are even less amenable to manipulation and recreation with, say, radiation techniques.

Many U.S. agencies and organizations support programs designed to preserve genetic resources. Among the defects of the present situation is the widespread tendency for many important issues to "fall between the cracks" of existing commitments, such that no one takes full responsibility for facing up to them. Specific aspects of this situation are discussed later in this report.

Several federal laws and regulations relate to preservation of germplasm resources. The most familiar of these is the United States Endangered Species Act, enacted in 1973, which provides for maintenance of an official list of threatened and endangered species and bans any construction activities (e.g., building dams) that would damage their native habitats. Enforcement of this law has forestalled several ecologically undesirable projects. On the other hand, other laws, such as USDA importation rules, albeit effective in preventing infection of U.S. livestock with certain disease organisms, have had the undesirable effect of hindering highly relevant importations for scientific research. Overall, the body of the law that deals with genetic resources is diverse and without central focus.

Out of concern for the increasing numbers of species listed as endangered or threatened, we have considered how best to promote the continued existence of all truly endangered species. After weighing all available measures for preserving endangered species under controlled conditions, we are repeatedly forced to the conclusion that the only reliable method is in the natural habitat. Knowledge required to preserve all of these organisms under artificial conditions is not available and, if it were, this approach would be prohibitively expensive. Preservation in nature does not require that we understand it, only an awareness that it must not be destroyed. When preservation in nature is not possible, intelligent zoo or garden maintenance, with encouragement of breeding potentiality, offers a possible alternative.

We can hope in the future to be much more effective in preserving existing germplasm resources. Saving the rich diversity of genetic material that has been provided by natural mutation and evolution can be achieved and is worth whatever effort may be required. It is critically important that the people of the United States recognize the long-term dangers inherent in loss of specific genes and of genetic diversity, recognize that diversity in germplasm is an essential national resource, and treat it as such.

2 NATURAL ECOSYSTEMS AND BIOLOGICAL DIVERSITY

Significant human impact on natural ecosystems is a comparatively recent event. Modern agriculture is our
greatest single ecosystem rearrangement, yet in the pre-
agricultural situation of just 10,000 years ago we were
hunter/gatherers playing a role in the ecosystem scarcely
different from other animal species. Clearly all of the
horticultural plants, domesticated animals, and useful
strains of microorganisms upon which we now depend are
the product of a very rapid and recent evolution under
human selection from wild relatives found in natural ecosystems around the world. These ecosystems, undisturbed
by agriculture, represent millions of years of evolutionary
change and adjustment to changing life forms and environmental conditions. All forms of life have evolved within
these natural systems; in turn, natural ecosystems have
evolved to support the greatest possible diversity of
living systems. A natural ecosystem with a diminished
diversity of living systems is thus an impoverished system
and any organism in a natural ecosystem with a contract-
ing genetic diversity is a threatened organism. It is in
the best interest of human society to see that the diversity
of natural ecosystems does not appreciably diminish.

BASIS OF DIVERSITY

Natural ecosystems depend on an enormous number of environmental and biological interactions and interdependencies,
and while many of them are highly fragile, yet they have
survived for millions of years without human intervention.
The welfare of each biological component within the community depends on the preservation of its genetic diversity,
which variability in turn requires an adequate population
size; the number of individuals needed to assure variability differs considerably from species to species.

5

To appreciate some of these complexities, consider the
relationships of higher plants and animals to the many
species of microorganisms with which they are associated.
Certain microorganisms are essential to the well-being
of a multicellular host; in many cases the host organism
has evolved such dependency on its microbial symbionts
that it cannot be regarded as an independent and self-
sufficient entity. Consider, for example, the majestic
Douglas firs of the Pacific Northwest and their dependence
on certain fungi to form root mycorrhizae. Without soil
fungi these trees simply will not grow. It is therefore
essential that means adopted for the preservation of one
plant also provide preservation for its symbiont. These
complex interactions are usually best preserved in intact
natural ecosystems.

At present, so few symbiotic relationships are understood in detail that it is simply not possible to decide,
in most cases, which members of the microbial flora of
plants and animals would need to be preserved separately
to provide for healthy stocks of the host organisms. A
substantive research effort would be required before such
decisions could be made. Techniques for the isolation,
cultivation, and preservation of many symbionts are not
developed as yet. Most have not even been accurately described or identified.

Free-living microorganisms that inhabit soil and water
are generally regarded as being ubiquitous. There does
not appear to be much danger that any of these species of
essential worldwide distribution face extinction. On the
other hand the microbes that live in close association with
plants and animals are in many cases highly adapted to this
association. These highly evolved microbial symbionts can
be expected to disappear when their host becomes extinct
or is removed from its native habitat.

Another category of interrelationships is that between
flowering plants and their pollinators. Adaptations are
frequently so precise as to require the presence of both
members of the partnership--plant and animal--for survival.
Many of these situations are still unknown while hundreds
have been discovered and intensively studied.

A commonly cited example of the importance of animal
pollination--in this case by a wasp--is the initial failure of the fig to set fruit when introduced into the Great
Valley of California in the last century. The trees grew
well and all signs indicated that the crop would succeed,
but the trees bore no fruit. What had occurred, of course,
is that when the tree was brought from its native habitat

along the Mediterranean the wild wasp pollinator was left
behind. In the absence of its pollinator, the specialized
flowers on the fig were nonfunctional. Only after research had uncovered the role of the wild wasp and it too
was introduced, was successful cultivation of figs possible
in California.

The survival of each species depends on the preservation
of the intraspecific variability of the gene pool, plus
its interrelationships with other organisms. No single
organism possesses more than a fraction of the genetic
variance of the species; thus populations are the smallest
unit holding the genetic potential of a species. The
maintenance of a species apart from its natural environment for significant periods of time is an uncertain means
of preserving germplasm because of the expenses involved
and because within a few generations the gene pool may
become modified as a result of artificial and systematic
selection pressures. The extreme examples are cultivated
plants and domesticated animals, which have been so altered
by selective breeding that they can no longer reproduce
and compete in the ecosystem from which their wild ances-
tors originated. This loss of variability in crop plants
has been further commented upon by Janzen (1973).

IMPACTS OF EXTINCTION

The loss of a species is an irreversible event terminating
a process that has taken thousands or even millions of
years of evolution (Prance and Elias, 1977). Replacement
of this biological diversity through the evolutionary
process will be extremely slow. The loss of a species
will inevitably affect the ecosystem within which it
existed. The ecological adjustments that are thus induced
are difficult and often impossible to predict and may
turn out to be deleterious to human welfare.

Man has had major impacts on natural ecosystems. These
impacts include the destruction of many species of large
mammals during the evolution of hunting and gathering
cultures (Martin, 1966) and the early destruction of vegetation as a consequence of agriculture and overgrazing
(Sears, 1935; Lowdermilk, 1953), which has been accelerated
by technological developments that came in the wake of the
industrial revolution and by the subsequent explosion of
the human population, due in large measure to improved
public health measures and to an agricultural abundance.

Certainly the greatest perturbation of natural ecosystems has been land clearing and the substitution of dense

pure stands (monocultures) of cultivated plants and/or
domesticated animals. In the past few decades, with rapid
increases in yield achieved by employing hybrid crop varieties and new farming techniques, our dependence on
future genetic improvements has increased rather than decreased. Over large areas of the globe the genetic
uniformity of a few varieties and races is displacing the
thousands of local gene combinations. This change has
its most conspicuous implications in the less-developed
parts of the world. The process represents a paradox in
social and economic development in that the product of
technology (breeding for yield and uniformity) displaces
the resource upon which the technology is based (genetic
diversity of locally adapted land races) (Wilkes, 1977).

Biological stability is in some measure achieved
through genetic diversity. For example, in a wild population there exists in individual plants a fairly wide variation in their ability to withstand cold, drought, disease,
insect damage, and other environmental variables. There
are several ways to maintain stability in a predator-prey
relationship, such as dispersing the prey at a low density
so the predator will only occasionally encounter it;
changing the prey from season to season; or ensuring the
prey has great genetic diversity. In this last case, the
predator will affect only a segment of the population.
In most nonindustrial agricultures, there is normally considerable diversity between fields of different cultivators,
as each maintains his own seed supplies. Suddenly, now,
we realize that genetic uniformity is beginning to sweep
around the world, as native varieties and breeds are
dropped in favor of introduced seed or breeding stock.
Quite literally the last vestige of the genetic heritage
of a millennium in a particular valley can disappear in
a single bowl of porridge. This genetic diversity was of
no great value until it was the last, then suddenly it
becomes immensely valuable.

Genes can be stored solely in living systems. Concern
about their loss stems from the irreplaceable nature of
genetic wealth. Once the systems are dead, the genes
they possessed can no longer be retrieved. Wild plants
can, of course, also be pushed to extinction by human
activities, but the process is usually slower because it
requires complete destruction. Indeed, although the
American prairie, even after a century of land clearing
and farming, is but a shadow of what it once was, much
of the flora still exists. This is not the case with the
native varieties and land races of cultivated plants,

domesticated animals, and fragile habitats under human usage. Besides undergoing the slow process of genetic erosion, these biological units have been known to disappear in a single year. For this the term "genetic wipeout" has been used (Harlan, 1975).

Examples of recent man-made perturbations of natural ecosystems include the destruction of tropical rain forest by land clearing for agriculture; the replacement of natural forest by tree monocultures; pollution of water and air by industrial waste material; the obliteration of wild areas by urban sprawl; the modification of water courses (for power, irrigation, and flood control); the drainage of wetlands; the use of broad-spectrum pesticides and heavy applications of fertilizers; overharvesting of natural populations (e.g., whales, anchovies, tigers, passenger pigeons, and black walnut trees); and the deliberate or inadvertent introduction of species to new areas (e.g., chestnut blight and Dutch elm disease in North America or various species of mammals and cacti to Australia (Ricklefs, 1973).

Despite these many impacts of the human population, some relatively undisturbed ecosystems remain intact, even though in possibly modified and/or depleted form. They serve as self-sustaining reservoirs for a vast number of species, each with its natural gene pool. Thus it seems obvious that an adequate number of these ecosystems must be identified and appropriately protected if the world's germplasm resources are to be preserved.

The irreversible effects of extinction are not fully understood and may prove to be deleterious. The preservation of species in natural areas may keep open the options of ecosystem restoration at a later time.

It may not be possible to save several examples of every habitat, or the natural home of every species. Making decisions as to which habitats or which species are dispensable is extraordinarily difficult, if not impossible. Who would have guessed the value of a certain strain of *Penicillium* on a rotted melon in the Peoria, Illinois, vegetable market before the discovery of antibiotics; of *Rauwolfia* before the discovery of the action of reserpine (a tranquilizing drug); of the giant squid before the discovery of giant neurons and their use in neurophysiological research; and of certain species of insects and fungi as biological pest control agents before they were found to be useful. In the absence of specific knowledge, the best defense is the one exploited by natural ecosystems: genetic diversity.

3 PRESERVATION OF NATURAL ECOSYSTEMS

A report by the Nature Conservancy (1975) for the U.S. Department of the Interior addresses the issue of preservation of natural diversity in considerable detail. It underlines the desirability of setting aside natural reserves now, before diversity is further reduced by unwise land and water use. It recognizes that when its natural habitat is not preserved, a given species is more likely to become extinct. It advocates that a coordinated system of ecological reserves be established within the United States. We would add, also, that long-term monitoring of such reserves would facilitate distinguishing among cyclic, stochastic, and man-made changes in the system and thereby enhance capability for making predictions with respect to natural ecosystems.

The conservation of wild habitats is by no means the only aspect of ecosystem preservation; one must also consider artificially managed habitats, with their subclimax environments, many of which support a highly desirable and diverse flora and fauna. For such habitats there must be a clearly defined goal as to the desired state of the ecosystem and an intimate understanding of the way in which the elements of the system relate to one another, so that management can be directed toward maintenance of the system.

In estimating the cost of establishing and maintaining natural and artificially managed ecosystems, it must be recognized that such a program will provide society with much of value beyond the one addressed here: the preservation of biological diversity. Few attempts to estimate these values have been made, but an appropriate example is a study of the Alcovy River in Georgia (Wharton, 1970), which showed that the bottomland swamp forest was substantially more valuable in an undisturbed state than if if it were to be channelized and drained for agriculture. These assets included flood control, water purification, timber production, wildlife and fish production, and recreational and educational uses. Natural ecosystems serve vital functions in the maintenance of environmental quality. Protection of watersheds, recharge of the water table, and improvement of air quality are examples. Some of these areas are subjected to varying degrees of human use, ranging from more passive types of recreation (hiking, camping, canoeing, mountain climbing, and photography) through more active forms that have greater impact on the biota (hunting, fishing, and traversing by various types of motorized vehicles), to such exploitative activities as grazing and timber harvest. As for exploitation, management of ecosystems often conveys distinct advantages over "wilderness" natural ecosystems. Seed collecting for use in forest tree breeding programs stands as an example.

It is clear that any attempt to preserve biological diversity must confront the many conflicts of interest that arise. Common examples are: competition for winter range between elk and livestock; pressure for predator control programs as against their impact on wolves and mountain lions; hunting pressures and tourist safety programs in relation to the grizzly bear; drainage of wetlands areas for farming at key points on migratory flyways and on breeding grounds of aquatic species; and impact of massive use of pesticides on such nontarget organisms as bees, birds, and fish.

An essential feature of a successful program for the preservation of ecosystems and endangered species must therefore be the development of a cohesive plan for each reserve. The plan must include the enunciation of objectives and formulation of mangement policies to ensure that the objectives are achieved.

TERRESTRIAL AND FRESHWATER ECOSYSTEMS

Terrestrial ecosystems in the United States have been repeatedly classified and inventoried. An early attempt was published in 1926 (Shelford, 1926) by the Committee for the Preservation of Natural Conditions created by the Ecological Society of America. The magnitude and complexity of the problem are enormous, particularly when one is concerned with the presence of the rare components. Küchler (1964) has developed a vegetation map for the entire United States that divides ecosystems into major

types. At the state level there are many natural-area
inventories, information about which is readily available
(Nature Conservancy, 1975).

Freshwater ecosystems--springs, streams, lakes, rivers,
and their associated wetlands--are threatened in many
ways. The threats take such forms as: pressures for use
of the water for domestic, agricultural, and industrial
purposes, which are particularly acute in the arid regions;
manipulation of water levels for flood control and reservoir impoundment; dredging for navigation; drainage for
agriculture; chemical and thermal pollution; solid-waste
disposal; filling for real-estate development; and mission-
oriented management for special interests (e.g., sport
fishing). Goodwin and Niering (1975) have summarized the
major human impacts on inland wetlands and have discussed
their classification.

Populations in these aquatic ecosystems are frequently
isolated because they cannot disperse widely, and a high
degree of endemism therefore develops, especially in arid
and semiarid regions. For example, endemic species of
desert pupfish are found only in certain springs along
the bed of the Amargosa River in Nevada (Miller, 1967).

The biota of aquatic habitats is not as well known as
that of most terrestrial ecosystems. Hence, the dimen-
sions of past and potential species extinction are less
clear. Furthermore, the classification of aquatic habitats
is not as well developed, although the U.S. Fish and
Wildlife Service has devised a system for categorizing
wetlands (Martin et al., 1953; Shaw and Fredine, 1956).
Preservation of freshwater organisms is especially difficult because information about them is inadequate and
because their habitats are so often exploited to fulfill
human economic and recreational requirements.

Any attempt to preserve freshwater organisms must take
into account watersheds within the natural-areas system,
as well as efforts directed to the restoration of water
quality in our lakes and streams.

The Nature Conservancy promotes natural heritage programs on a state-by-state basis and guides their development. These programs are based on detailed inventories
of the biota and emphasize rare species and plant associations. Further, the Conservancy has developed open-ended
computer programs for compiling information that will
permit the eventual development of a national data bank
(Jenkins, 1977; Moyseenko et al., 1977). The computer work
has thus far been funded by modest grants from the National
Science Foundation and private sources. Data gathering
has been funded from state heritage program budgets.

MARINE ECOSYSTEMS

Species in the marine environment are by no means fully
inventoried, although some communities, such as those
occupying the rocky intertidal zone along the coasts of
Europe and North America, are comparatively well documented.
But the most diverse areas--the coral reefs and the deep
sea--are very poorly known. In general, the marine environment has received much less attention than have the
terrestrial and freshwater environments.
  Diversity of species is, of course, the result of a
continuing process of speciation and extinction. A
general understanding of speciation has been arrived at
for only a few groups of animals. It seems likely, however, that there are significant differences between groups
of animals in the way in which speciation occurs, and
great differences between environmental regimes in the
rate at which species are formed and become extinct. It
is generally believed that species diversity confers stability on ecosystems; recent theoretical analysis (May
1973) suggests that precisely the opposite may at times
be true. It may well be that diversity does not always
confer stability on a system, that the most diverse ecosystems are the most vulnerable to disturbance. In other
words, such highly diverse ecosystems as the tropical
coral reef and the deep-sea benthos may rely for their
existence on environmental stability. Thus the very systems that are the least described and whose dynamics are
least understood may be the most fragile of all.
  Estimates of genetic variability indicate that species
of marine invertebrates are in general highly polymorphic
(Selander, 1976). Current evidence suggests, however, that
variability is low in certain large marine crustaceans
(lobsters and crabs) (Tracey and Nelson, 1975) and some
estuarine species (Gooch, 1975).
  One method for deriving comparative estimates of genetic
variability is by electrophoretic survey of soluble products of structural genes. It is not known what proportion
of the genome codes these proteins or what proportion codes
the regulatory genes responsible for gene activation and
expression. There are many instances in which morphological differences between species are not closely coupled
with evolutionary divergence. For example, inshore and
offshore populations of the well-known species of diatom
Thalassiosira pseudonana are very different genetically,
although similar morphologically (Murphy and Guillard,
1976). Comparable observations have been made on summer

and winter populations of another diatom, *Skeletonema costatum* (J. C. Gallagher, personal communication). *Capitella capitata,* a cosmopolitan polychaete worm used as a pollution indicator, is now thought to be a complex of six sibling species that are morphologically similar, show almost no genetic overlap, and have sharply different life history characteristics.

Much of the genetic diversity in marine species resides in latitudinally separated populations whose systematic interconnections are unknown. An improved understanding of the role of planktonic larvae in linking discontinuous populations would help in elucidating the factors controlling the enormous year-to-year variations in recruitment of young that are so apparent in many commercial species of fish and shellfish. It is not now possible to distinguish between long-term cycles in the marine environment and unidirectional trends stemming from increasing effects of human activity. This distinction can be made only by instituting long-term studies of selected marine communities.

Although the major geographical boundaries in the oceans are well defined, to determine the degree of geographical isolation between populations in the marine environment is much more difficult than for terrestrial or freshwater systems. In many cases the presence of a particular species or community in a given body of water is the most informative indication available as to the history and physical and chemical properties of that water mass. Land-based and marine ecosystems differ greatly in the quality and quantity of information available at various levels of classification and inventory. Other important differences have to do with procedures for establishing and managing natural areas. Thus far, as a consequence, very few marine preserves have been designated. These differences may be partly due to the relatively recent recognition accorded to the desirability of conserving and preserving marine resources. The degree of recency can be appreciated by recalling that the national forest system was established in 1905 and that the first federal natural-areas site was set aside in the Coronado National Forest in 1927 (AIBS, 1974). By contrast, it was not until 1945 that President Truman proclaimed the natural resources of the continental shelf to be under the jurisdiction of the United States, and it was not until 1972 that Congress established an estuarine sanctuaries program as part of the Coastal Zone Management Act.

For several reasons, proposals for establishing an

international system of reserves, with the primary objective of conserving genetic diversity, stop at the coastline, except for near-shore marine ecosystems in the Everglades National Park, Florida, and the Virgin Islands National Park. Plans of this nature were included in UNESCO's Man in the Biosphere (MAB) program and in the U.S.-U.S.S.R. Environmental Agreement (Franklin, 1977).

It may, indeed, be premature to expect comprehensive plans applicable to estuarine and marine environments. After all, not until 1974 did the Office of Biological Services, U.S. Fish and Wildlife Service, begin a major effort to revise and expand its classification of wetlands and aquatic habitats (including marine) with a view to producing a new inventory of wetlands. Odum et al. (1974) developed one classification of coastal ecological systems and subsystems, but a comprehensive assessment of estuarine and marine ecosystem preservation needs to be undertaken. The International Biological Program (IBP) was initially designed to obtain more information about the relation between biological productivity and human welfare, and in the United States included a section entitled Conservation of Ecosystems (IBP/CE), which in turn had an estuarine marine task force that developed a classification system for aquatic and related environments in the United States (Darnell et al., 1974). Marine areas already designated for protection comprise primarily coastal lands. Adjacent and offshore waters were either not included or the administering body did not have regulatory authority over activities in those waters. The IBP terrestrial and freshwater task force did, however, manage to compile an inventory and assess the adequacy of representation of various ecosystem types (Darnell et al., 1974). More recent attempts have been made to classify marine and estuarine habitats (Ross, 1974; Ray, 1975; and Kifer, 1975).

A number of baseline studies, now being conducted in the marine environment, should contribute to classification and inventory of marine ecosystems and lead to a deeper understanding of the dynamics of those systems. In the past, however, the expenditure of large sums of money on broad-scale sampling programs has not brought about such an understanding. Too often a piece of information crucially needed to make the whole set of data meaningful has been lacking, perhaps because there was no testable hypothesis guiding the sampling strategy. Funding policies should be such as to facilitate long-term studies by competent scientists and by institutions of established excellence, so that trends, periodicities, autocorrelations,

and random fluctuations can be examined. Moreover, better
communication should be established between federal agencies involved in environmental problems and ecologists
engaged in fundamental research, such that the design of
long-term data collection may be improved and existing
data sets made more readily available.

It is clear that more effort needs to be devoted to
gathering basic data on estuarine and marine ecosystems,
especially at the classification and inventory level,
and to long-term monitoring. Furthermore, legislation
to preserve natural areas, and agency regulations and
guidelines, should capitalize upon this information as
it becomes available.

Long-term monitoring should be done on both unspoiled
and perturbed areas so that predictions can be developed
as to the probable effects of disturbance on different
kinds of communities, ranging from the least to the most
diverse. This approach should throw some light on the
vexing question of the optimal size and number of areas
to be designated for preservation (Diamond, 1976;
Simberloff and Abele, 1976; Terborgh, 1976; Whitcomb et al.,
1976). Ecologists seems to agree that different species
show different susceptibility to extinction. A strategy
for preserving large, long-lived species at the top of
the trophic ladder will obviously be different from one
designed to protect relatively opportunistic species with
high dispersal ability and short generation times. In
general, the size of ecological preserves should be maxi-
mized, but a predictive theory for the fate of individual
populations in, say, the marine environment requires much
more information on larval dispersal, recruitment, and the
frequency of local extinctions. It is likely that successful management will have more to do with biology of
local populations than with efforts to encompass the
total area.

THE PRESERVATION OF OBSCURE AND UNKNOWN ORGANISMS AND
ENDANGERED OR THREATENED SPECIES

The preservation of habitats is not only the most convenient means of maintaining gene pools of many species
of unknown value; it is the only means of maintaining the
germplasm of unknown organisms. Because the larger and
more conspicuous plants and animals, particularly in North
America, have long since been catalogued, one may be in-
clined to think that the inventory of nature is nearly

complete. But many of the more obscure organisms have yet
to be systematically examined and can be classified only
by general type. What few formal surveys have been conducted divide the paramecia, for example, into one of the
dozen or more species complexes such as *Paramecium aurelia,*
P. multimicronucleatum, or *P. caudatum.* Yet the likelihood is that each of these species complexes, if carefully
examined, would be found to include several or many species
that could be distinguished with the investment of appropriate effort (Sonneborn, 1975; Nanney and McCoy, 1976).
This "under-classification" may be characteristic of the
protists in general, of some groups of "uninteresting"
plants, and of some of the less-conspicuous invertebrates.
In many cases evidence for genetic discontinuity has never
been sought. When diverse organisms are lumped together
as a species, instead of being treated as a species complex, a particular component may be destroyed without ever
being recognized or dignified with a Latin bionomial. The
only possible strategy for the preservation of such species
is the maintenance of a diversified system of natural
habitats.

A special obstacle to preserving endangered species per-
tains with respect to organisms whose genetic economy is
based on the exploitation of local habitats by genetic
specialization (Sonneborn, 1957; Nyberg, 1973). Coloniza-
tion and inbreeding produce many unique local populations,
often largely or completely isolated from other related
populations; hence they "speciate" freely into many local
gene pools. Other organisms reflect the strategy of
physiological responsiveness to adjust to environmental
variables. When they outbreed, these organisms differ
little in their genetic components over broad geographical
areas; they rarely speciate, and their preservation is
relatively easy.

The difficulty stems from the inflexibility of current
legislation concerning endangered species. The loss of
a "local" species of the *Paramecium aurelia* complex would
probably be a far less significant event than the loss of
a widely distributed species of the *P. bursaria* complex,
but regulations do not now permit such distinctions. If
one knew the propensity of certain kinds of organisms to
speciate locally, one could probably discover a unique
species in any habitat whatsoever chosen for preservation.
Because relatively few of the local species have in fact
been identified and named, the demonstration of a previously
unknown species in an area perforce gives that area an
aura of "importance." Although the tactic of using

endangered species to block public works is often pragmatically feasible on a short-term basis, it is ultimately
self-defeating.

The legislation must eventually be refined sufficiently
to permit a more realistic assessment of the competing
values involved. A realistic assessment must include an
understanding of species diversity and species multiplicity
in outbreeding and inbreeding species. A change in emphasis
from protecting single endangered species to conservation
of the total diversity in representative ecosystems would
in part mitigate current weaknesses of the Endangered
Species Act.

The Endangered Species Act has provided a strong legal
means by which man's impact on the natural environment can
be halted or reversed. In many cases where cogent arguments about postulated undesirable effects of environmental
modification on whole ecosystems have failed to persuade,
the Endangered Species Act has served as a means of last
resort. Many examples could be cited where preservation
of the critical habitat of a single nonresource species
has halted multimillion-dollar projects of long standing.
For example, early in 1977 the Tennessee Valley Authority
was enjoined from further construction of the almost completed Tellico Dam on the gounds that completion of the
project would destroy the habitat of the snail darter
(*Percina tanasi*), a small fish found only in the Little
Tennessee River. In delivering the opinion for the United
States Court of Appeals for the Sixth Circuit, Judge
Anthony J. Celebrezze and two other justices wrote:
"Whether the dam is 50 percent or 90 percent completed
is irrelevant in calculating the social and scientific
cost attributable to the disappearance of a unique form
of life. Courts are ill-equipped to calculate how many
dollars must be invested before the value of a dam exceeds
that of the endangered species. Our responsibility under
the Endangered Species Act is merely to preserve the status
quo where endangered species are threatened, thereby
guaranteeing the legislative or executive branches sufficient opportunity to grapple with the alternatives."
It could be added that it is not just the courts that have
failed to develop the means adequately to weigh the value
of nonresource species; the scientific community is similarly lacking in criteria.

To distinguish between threatened and endangered species,
it is necessary to know the size of the population, to
assess the change in size with time, and to have an understanding of the life cycle, population dynamics, ecological

relationships, and habitat requirements of each species within the ecosystem. Habitat requirements for migratory birds along the length of their flyway pose special international problems. The status of a species can change swiftly as a result of human actions. Thus, if an exotic species is introduced or allowed to escape, the species may have drastic effects on the native biota; or a species may be overexploited; or an insecticide may be concentrated in predators at the top of the food chain. For example, very efficient harvesting of Antarctic krill might disrupt the ecology of that Antarctic ecosystem and endanger less common components of the system dependent on the krill as food (Shapley, 1977).

The total number of endangered plant species is unknown. Because of the relatively small number of specialists, it is not now possible to determine whether a given organism is really endangered as a species or only as regards one ecotype (e.g., Monterey pine). The conservation of small shrubs and herbaceous species relies more on guesswork than genuine understanding of their biology. Data are needed not only on their ranges, but also on their "requirements." Weber and Johnston (1976) are among those who have compiled lists of rare and endangered plant species, but the distinction between a species of plant and a variety has brought about discrepancies between various listings.

An examination of the list of endangered and threatened wildlife and plant species in the *Federal Register* (Vol. 41, No. 208, 1976) indicates that certain animal groups are well represented and that others are wholly absent. The gaps are particularly conspicuous when one looks for estuarine and marine species. Apart from a few species of large, long-lived animals (e.g., whales, manatees, crocodiles, and the short-nose sturgeon), species from the marine environment are almost entirely missing, and marine invertebrates are not represented at all. This is due in large part to comparative ignorance of the status of marine species and in some measure, no doubt, to the difficulty of finding advocates for small, insignificant-looking organisms.

Lists of endangered species and their habitats must be founded on an adequate data base. There should be a world-wide inventory of all types of organisms, of which a comprehensive national inventory for the United States would be an important part. Conflict with public works projects might well be avoided in the future if the habitats of endangered species had been thoroughly catalogued.

Were such a change in emphasis to occur, there would still be a number of endangered species, particularly certain large species of mammals and birds that will claim special rights to protection, either because they are of great aesthetic or economic importance or because they are a valuable research resource. The status of such species is considered by Zisweiler (1967), who has documented the accelerating rates of extinction that can be related directly or indirectly to the effects of human populations. The Red Data Books of the International Union for Conservation of Nature and Natural Resources (IUCN) provide an extensive analysis of mammals, birds, reptiles, amphibia, and fish in various degrees of endangerment. For such endangered species, guardianship in zoos or aquaria may be the only way of preventing their extinction; for other species, careful restoration of the natural habitat and a relaxation of the pressures previously leading to endangerment may be successful (e.g., the Hawaiian nene goose) (Martin, 1975); for others (e.g., the whales), extinction can only be prevented by the implementation of international agreements.

COOPERATIVE EFFORTS TO EXPEDITE CONSERVATION OF ECOSYSTEMS IN THE UNITED STATES

Public funds must be devoted to projects involving publicly owned areas. There is a need for coordination between action addressed to these areas and a number of other types of land holdings.

Many conservation organizations and institutions of higher learning own land characterized by managed or natural ecosystems. Some of these areas are being used for conservation programs; others survive through benign neglect. If all such areas are taken into consideration, there is very likely a larger capability for conservation than is generally recognized. Small-scale conservation efforts need to be publicized and help given to institutions, where needed, to preserve some segment of the ecosystem. Organizations involved in this kind of activity need to be made aware of the importance of their work and given an opportunity to participate in a network that would help to conserve habitats throughout the country.

In many cases the ecosystem harboring a given species may already be destroyed or unsafe. Banking in zoos or specific reserves has been instituted for some such species (e.g., Przewalski's horse and Arabian oryx) with the hope that they may ultimately be restored to their natural habitat. Rigorous restoration of the habitat and freedom from former pressures must be guaranteed before such repatriation is attempted. A comparable situation exists for many plant species, suggesting that botanical gardens and arboreta can play a role in preserving plant taxa that might otherwise disappear.

The effects of the loss of species and habitats are not fully known, and an argument for preserving diversity is often our ignorance of what would ensue if we neglect to do so. It would be unrealistic to expect to be able to preserve every living organism, no matter the cost, and a system of evaluation to facilitate decisions as to what should be actively preserved is urgently required. Further knowledge of species interactions would help to predict the consequence of extinction of one organism in an ecosystem. Until such knowledge has been developed it would seem prudent to make a major commitment to the identification and protection of as many ecosystems and endangered species as possible.

4 PRESERVATION OF ECONOMICALLY IMPORTANT PLANTS AND ANIMALS

THE VALUE OF COLLECTIONS

For centuries a wide variety of organisms have been collected and maintained as sources of food, clothing, shelter,
transport, medicine, and research materials. Aside from
the obvious convenience of having the organisms available
when needed, collections make it possible to select for
varieties that are better adapted to human needs than those
found in nature. As a corollary to the selection process,
many collections have been developed through the activities
of investigators interested in the basic biology of the
organisms, as such.
    Many useful strains that have been developed through
selection exist only in specialized collections, and such
strains, which often represent a very large investment,
are virtually irreplaceable if lost. However, the selection process can itself result in serious loss of germplasm.
Selection for particular traits at the expense of others
very often leads to a narrowing of genetic diversity in
the species. The reduction in genotypes is particularly
likely to occur as techniques for mass rearing are adopted
and culture conditions become more uniform through artificial control. The discarding of all strains except
those best adapted to the conditions of domestication becomes a serious threat to the maintenance of the organism
if environmental conditions change, if susceptibility to
predators changes, if parasites or pests develop, or if
there is need to select for different sets of characteristics. Limiting the available types in a collection can
be particularly serious if human or other encroachment on
natural habitats has meanwhile depleted species diversity
or has eliminated the species from natural communities.

22

ROLE OF PLANT GENETIC RESOURCES IN AGRICULTURE

In cultivated crops, germplasm resources are required to provide the genetic diversity needed to ensure future production. Plant genetic resources include wild species related to the cultivated crops, uncultivated forms of the cultivated species, folk varieties (or land races), obsolete and current cultivars, useful mutants, and stocks with improved combinations of genes developed as a consequence of research. This array of genetic diversity is essential to meet the constantly changing problems imposed by consumer needs, agricultural technology, environmental changes, pests, economic conditions, and other factors.

In the United States, responsibility for crop improvement through plant breeding is shared by the federal government, state governments, commercial firms, and foundations. Between 450 and 500 new cultivars are released each year. Most represent minor genetic advances and "fine-tuned" adjustments to changes in production, harvesting, processing, and marketing procedures.

Until about the beginning of this century, farmers commonly put aside plants from each crop for use in propagating subsequent crops. The saved plants were those judged to be superior. This practice has led to occasional hybridization (sometimes intentionally, sometimes by chance) and to many "folk" varieties. Substantial genetic variability existed within and among these cultivated varieties. Moreover, that part of a species not chosen for cultivation generally survived in nature, because its natural habitat had not been destroyed by the pressure of human population or by agricultural technology.

Early in this century the circumstances that existed for so long began to change; they continue to change, and rapidly. Professional plant breeding, generated by the rediscovery of Mendel's laws and the development of the chromosome theory of heredity, began 60 or 70 years ago. Application of these scientific principles led to modern crop varieties, which are selected for uniformity, yield, quality, and adaptation to specific environments. Selective breeding programs and the adoption of these superior stocks derived therefrom have brought about remarkable increases in agricultural productivity. Unfortunately, they have also led to the abandonment of many old folk varieties and land races, thus accelerating the erosion of plant genetic resources. In addition, the world population exploded from something over a billion to 4 billion people. The industrial revolution, coupled with population pressure, disturbed the natural habitats of many species, and as a consequence the genetic resources resident in wild and cultivated plants shrank. Important crops began to rest on a narrowed germplasm base, because farmers no longer saved their seed and, instead, obtained higher-yielding strains from a limited number of commercial suppliers.

NEED FOR A BROAD PLANT GERMPLASM BASE

Agricultural leaders in this country and abroad now recognize that genetic variability of crops is shrinking and that valuable plant genetic resources are thereby being lost. We cite now three examples of the dangers of a narrowed base; two deal with crises that were met in the past, the other with a current problem.

Grapes are among the economically more important fruit crops in the world. In the latter half of the nineteenth century the wine industry of France (and other European countries) was virtually destroyed by the ravages of an insect, *Phylloxera*, that attacked the roots of the vines. In Europe, vines were of but one species, *Vitis vinifera*, thus making all plantings on the continent equally susceptible to attack. By contrast, approximately 30 species are recognized in America, and it was known that the roots of some or all of these were resistant to *Phylloxera*. They did not however produce wine of quality equal to the European species. The numerous varieties of European grapes had been developed through centuries of cultivation and selective breeding from the gene pool of the one species, a process that could not be repeated rapidly with the American vines. The problem was solved, and the threatened demise of the industry avoided, by grafting the French vines onto selected American rootstocks, on which they still grow today. Similar combinations of highly selected vines grafted onto native American species stocks are also widely used in California. But these very native wild vines of North America are now being seriously reduced by the inroads of man.

A current situation that calls for diverse germplasm resources has to do with the reclamation of lands in the western United States that have been strip mined. In these semiarid regions the reclamation of strip-mine tailings is extremely difficult. Most of the native species best adapted to the natural conditions of these regions are not good colonizers of disturbed habitats. Some of the

most promising species as candidates for introduction come
from other regions and include weeds from other parts of
the world. Yet reclamation with plants of this type carries appreciable risk because some of them may prove
troublesome in the new environment. The solution to this
complex environmental problem will require intensive research. It will require the testing of many species from
many localities.

As a final example, consider the Southern corn leaf
blight epidemic of 1970, which brought a new sense of
urgency to the situation when losses at harvest reached
50 percent in some states and 15 percent nationally.
This threat to a major crop created so much alarm that
the National Research Council appointed a Committee on
Genetic Vulnerability of Major Crops to examine the
cause of the epidemic, the vulnerability of our crops to
attack by pests and pathogens, and possible measures to
hold losses to low levels and reduce the likelihood of
epidemics. The report (Committee on Genetic Vulnerability
of Major Crops, 1972) includes this statement:

> Two points are clear: (a) vulnerability stems from
> genetic uniformity; and (b) some American crops are
> on this basis highly vulnerable. This disturbing
> uniformity is not due to chance alone. The forces
> that produced it are powerful and they are varied.
> They pose a severe dilemma for the sciences that
> society holds responsible for its agriculture. How
> can society have the uniformity it demands without
> the hazards of epidemics to the crops that an ex-
> panding population must have?

In partial answer to the above question, the Secretary
of Agriculture in 1975 established the National Plant
Genetics Resources Board to advise the Secretary on national needs for the assembly, description, maintenance,
and effective utilization of living resources in plant
improvement programs. The Board is preparing an analysis
of the status of crop germplasm resources in the United
States.

In 1976 the Board established liaison with the International Board for Plant Genetic Resources, which was
created in 1974 by the Consultative Group on International
Agricultural Research. The mission of the International
Board is to ensure the conservation of genetic variability
in economic species of plants to be used by plant breeders
and by research workers to promote improvement of cultivated

plants and of agriculture itself. To this end the Board expects to develop collaboration among the members of a global network of institutions active in the exploration, collection, conservation, documentation, and use of plant genetic resources.

CATEGORIES FOR CROP GERMPLASM MAINTENANCE

Plant germplasm, in the broadest sense, includes all living plants capable of reproduction. Most species, particularly the uncultivated ones, have survived without any direct aid from human beings. This category (uncultivated species) includes wild relatives of the crop species. Habitat disturbance reduces the ability of the natural ecosystems to provide plant materials that may be needed in the future. Appropriate action is needed to preserve natural ecosystems.

A second category includes folk varieties, "dooryard" plants, and land races that merit some protection. They are in the care of small farmers, horticulturists, and gardeners in all parts of the world. No one has any inventory or fixed responsibility for them. They are part of the cultivated ecosystem.

A third category includes plants that have been assembled by scientists or amateur botanists. It is not unusual for individual scientists, employees of a research station, or employees of commercial seed companies to accumulate germplasm collections well beyond their immediate needs. Most are willing to share their stocks with others in an informal, uncoordinated system. Although this system is valuable, it is difficult to ascertain what stocks are available. It is vulnerable to losses as people retire and as administrators reevaluate priorities.

Most plant scientists feel that these three categories are not wholly adequate to meet crop germplasm needs. A fourth category, known as the National Plant Germplasm System, has evolved over the years.

THE NATIONAL PLANT GERMPLASM SYSTEM

The National Plant Germplasm System is a part of the USDA (ARS, 1977); the National Plant Germplasm Committee advises on its organization, administration, and operation. Although the system is fully operational and well established, the Committee has identified the following activities as meriting the earliest practicable attention.

Establishment of Repositories for Clonally Propagated
Plants

Species that can be preserved by seeds are more easily
managed than are those that must be preserved vegetatively.
Some species (e.g., potatoes) can be preserved as seeds;
however, the genotypes are so heterozygous that clonal
propagation is also advisable for some stocks so that
especially valuable genotypes will not be lost. Still
other species must be clonally propagated because the
plants do not reproduce by seeds. Compared with the seed
crops, the conservation of clonally propagating ones is
relatively uncoordinated and inadequately supported. The
National Plant Germplasm Committee recognized that more
attention must be given to clonally propagated plants. A
start has been made by developing a national plan for
maintenance of fruit and nut crops.

Establishment of a Tropical Facility

A facility for conducting research of special relevance to
the tropics is needed. It should be in a latitude that
provides the length of day required for flowering, in the
field, of short-day or photoperiodic stocks. It should
include a winter nursery where U.S. research workers can
grow a second generation each year.

Funding Selected Curators

There are many individuals who maintain germplasm materials
beyond their immediate needs and thus represent potential
informal curators. Most are willing to share materials,
but have had to support their curatorial activity out of
research funds and, in an era of declining research budgets,
this function has usually been the first to suffer. In
the past few years, although funds have been provided to
selected curators, most have not had any additional support.
The National Plant Germplasm Committee found that many
curators recognize their national responsibility for the
germplasm in their care, which responsibility may have
been channeled through regional or interregional projects
or other cooperative agreements. But many other curators
are holding germplasm because of personal interest or because they are working for an experiment station that
undertakes to keep it as an adjunct to its research on

some crop. They have in fact no formal responsibility to anyone and may dispose of the germplasm as they please. The Committee is attempting to determine who wants to serve as "formal" curators, i.e., persons and stations willing to assume responsibility for collection, maintenance, preliminary evaluation, recordkeeping or documentation, and distribution to qualified users.

The Committee will recommend specific funding for these curators.

Identification of Gaps in Major Collections

Support for foreign and domestic plant explorations comes from various sources, i.e., state, federal, and private funds. One account administered by the Agricultural Research Service, USDA, is used exclusively to support plant-collecting expeditions, about six per year. In 1976, for example, three foreign plant explorations and two domestic explorations were carried out. Germplasm of several species of cotton was collected in remote areas of Honduras, Nicaragua, and Mexico. Ecological data were also gathered on the boll weevil and other cotton insects. Primitive tomato varieties were collected in Panama, Costa Rica, Nicaragua, Honduras, and El Salvador. Citrus relatives for possible use as rootstocks and for breeding were collected in Australia and New Guinea. Some of the grapes native to the northeastern United States were collected and are being evaluated. Bitter gourd, a potential new oilseed crop for the arid Southwest, was collected from throughout the southern Rocky Mountains and Great Plains; the seeds will provide the basis for crop-development programs in several states. Other recent collections include grasses in Russia, legumes from Italy and Greece, potatoes and peanuts from South America, and sunflowers and pecans domestically.

Computerization of Information

Various elements of the National Plant Germplasm System have turned to computers as an aid in managing data on collections. For example, computers are being used by the USDA at the National Seed Storage Laboratory, Principal Plant Introduction Office, Western Regional Plant Introduction Station, and others. In 1976 a contract was let by USDA to assess various ways to develop a national

system for storing and retrieving data, and suggestions
sought from users and keepers of germplasm in an effort to
design a system of optimum use to both groups.
Particularly useful would be expanded data on such
characteristics of the stocks as quality, pest resistance,
stress tolerance, growth habit, region of adaptation, and
others.

Research on Techniques for Germplasm Maintenance

Long-term storage of germplasm is expensive. Research
is needed to make storage more efficient, more secure,
and less expensive. For example, research on the effect
of temperature, moisture, atmosphere, and method of con-
tainment on longevity of seeds could lead to improvement
in storage facilities and procedures. Too little is known
of the cryobiology of seed, pollen, and tissue storage.
Clonally propagated plants present a special challenge.
Germplasm of fruit crops is maintained in orchards through
whole-tree maintenance practices that require much land,
labor, and money. If cells or tissues could be dependably
stored and regenerated into whole plants, germplasm maintenance would be significantly less expensive. Much
research will be required to make such procedures sufficiently secure to replace traditional methods.

INDIGENOUS SUBSISTENCE AGRICULTURE

Activities such as the National Plant Germplasm System
are directed towards the agricultural needs of developed
economies. The germplasm situation in countries relying
on indigenous subsistence agriculture are often very
different.
Subsistence agriculture depends on a reliable annual
yield of crops, as much as possible irrespective of
weather and soil conditions. This is in contrast with
modern agriculture, which is geared to maximum yield of
genetically uniform select cultivars that respond to
definable water and nutrient levels. Native cultigens
may be nutritionally superior and tolerate a greater
diversity of environmental fluctuations. Analytical study
of indigenous agricultural systems from ecological, genetic,
and nutritional viewpoints would be valuable in assessing
the need for conservation of this germplasm. If such
systems appear nutritionally and ecologically sound for

a given region, their survival should be encouraged,
particularly where monoculture and the adoption of "green
revolution" varieties is costly and may lead to meager
returns because of uncertain soil and climatic conditions.

Any effort to incorporate indigenous subsistence agricultural systems into a genetic resources program must
rest on the principle that genetic diversity is not only
a function of the plants (past and present cultigens and
progenitors) but also of the human element, which provides
opportunity for expression of these variations and exerts
selection forces to maintain genetic heterogeneity. The
success of such a two-component process is a matter of
survival for subsistence farmers. The imperative for
assuring reliable yield on an annual basis in regions of
poor and unpredictable growing conditions mandates the
maintenance of very wide genetic variation in the food
plants that are cultivated.

The impressive genetic diversity available in the food
plants of the less-developed countries is now subject to
the same pressures that have so drastically reduced diversity in the developed countries. Road-building, urban
construction, and large-scale cultivation are reducing the
land areas that contain the folk varieties of subsistence
crops and their wild progenitors. In the case of some
crops (e.g., wheat, *Triticum*), the introduction of the new
cultivars produced in crop-breeding programs for large-
scale production is eliminating many of the folk varieties
or land races. It is well established that the genetic
diversity of such food crops as white potatoes (*Solanum*),
tomatoes (*Lycopersicon*), and sugarcane (*Saccharum*) has
been severely reduced.

Only recently have efforts been made to conserve the
genetic resources of the major crops characteristic of
tropical agriculture. There are still no efforts to
conserve those of lesser significance, such as the fruit
and nut crops, of which there are a very large number.
Vegetables, including many legumes, that are consumed by
local peoples in the tropics are very poorly known. Even
in the case of *Zea mays* (corn), for which germplasm has
been collected extensively since 1943, we do not have as
complete a collection as we should, and conditions for
storage and rejuvenation are often far from satisfactory.
This is due in part to failure to maintain valuable collections assembled within the past 35 years. They must
now be reassembled, if that is still possible.

EXPORT CROPS OF THE LESSER-DEVELOPED COUNTRIES

A wide variety of export crops are produced on a large
scale in the developing countries. These include tea,
coffee, rubber, oil palm, cacao, teak, and other forest
products. Major exploitation of some of these items is
having drastic and deplorable impact on worldwide germplasm resources. The germplasm base of the crops thus
exploited is being reduced through the planting of vast
areas with single types, such that the enormous native
diversity initially characteristic of these areas is
being rapidly destroyed. The developed countries benefit
from these crops in the short run, but their long-range
interests are endangered by current practices. The developed countries should take some responsibility for
curbing these inroads on the world's resources.
    Rubber is an important export crop in many parts of
the world but cannot be planted extensively in the Western
Hemisphere because this is the native home of the plant
(*Hevea brasiliensis*) and of a serious fungus pathogen of
its leaves. Asian tropical areas are near peak production,
and the introduction of the South American leaf disease
could destroy this productivity. Because petroleum products are the starting material for synthetic rubber, it
is probable that we shall become increasingly dependent
on this crop. In any event, for some uses synthetic
rubber is inferior to that from *Hevea*. It is therefore
of special importance that a pool of diverse germplasm
for this species be maintained as a resource for use in
responding to future needs.

FOREST TREE GERMPLASM

Genetic diversity is fully as important for forestry
breeding programs as it is for agricultural crops. If
forestry operations are uncontrolled and ill-managed, and
natural forest gene pools indiscriminately mixed, selected
natural gene pools will be impoverished or even eliminated
(Yeatman, 1972; Maini, 1973; Barber and Krugman, 1974).
Forestry practices are becoming more intensive, and very
few stands will remain untouched after a few more years.
Native forests are often replaced by exotics or by non-
local reforestation stocks. Thus we should try to maintain ancestral types in order to ensure that a broad
pertinent genetic base is available for future selections.
    The problems associated with maintaining a genetic base

for forestry are not identical to those for general
agriculture. Each tree encounters many environmental
fluctuations during its long life span. Widespread species
tend to be more genetically variable than are ones with
restricted range, because of the greater diversity of
environments encountered. Races of a species growing in
different climatic regions may differ in their adaptation
to environmental factors, but the limiting factors may
differ from those governing cohabitation species
(Callaham, 1970). Often there is little information
about the most suitable seed source to use for a given
location; hence, by default, local seed sources are
employed.

To prevent loss of the original genetic base, strate-
gies for maintaining a reliable and varied genetic reservoir for future improvement should be developed, and
standards to gauge progress in genetic improvement and
perpetuation of large and small populations for future
mass seed production should be established (Yeatman, 1972;
Maini *et al.*, 1975).

Protection of the gene pool of comparatively long-lived
trees confronts fewer obstacles than does protection of
annual crop plants. In addition, U.S. foresters, with a
few exceptions, deal with wild populations of native
germplasm. Difficulties that must be faced include
preservation of diversity in areas of intensive management
and the risk of losing germplasm from species with dis-
junct distributions, such as isolated stands and outliers.
Some of the methods commonly used to maintain and protect
forest resources include:

● Seeds, pollen, and tissue cultures may be maintained.
Because not all material can be preserved, this approach
necessitates a decision as to the future desirability of
the varieties maintained.

● Selected stands may be set aside to preserve material
considered to be essential. If this is not possible,
plantations may be developed.

● Seed orchards and arboreta plantings are useful in
maintaining genetic selections, especially those related
to commercial forestry.

● Natural areas, national parks, and primitive and
wilderness areas provide significant reservoirs of genetic
diversity for forestry. They are, however, rarely established or managed for the express purpose of maintaining
a broad genetic resource for forestry. Regulations as to
the use of these areas often prohibit disturbance and

commercial activities, and therefore preclude mass seed collections for production forestry or related activities aimed at restoring damaged forest ecosystems.

Special gene pool centers for forest genetic reserves should be established. They should be representative of gene pools in areas where consumptive forestry is or will be practiced or where other pressures threaten the diversity. They should be large enough to contain the full range of biological and environmental diversity, to permit mass seed collections, and to minimize the hazard of contamination by foreign pollen.

Many current forest conservation programs include attempts to preserve the ecosystem in an unchanging state (e.g., by controlling forest fires). This may not always be desirable. Shade-intolerant species, important components of wood and fiber supplies, are often at a disadvantage in such undisturbed forest conditions. Management practices for genetic reserves should take into consideration whether it is advisable to aim for preserving an unchanging state.

DRUG PLANTS

Medicinal plants of importance fall into several categories:

- Those that yield pure chemical compounds of proven worth in the treatment of disease
- Those that are used in a crude or refined form and are therapeutically effective
- Those that yield chemical compounds of importance as starting materials for the semisynthetic production of useful drugs
- Those that yield extracts or compounds that in themselves have no appreciable medicinal value but may be necessary for the prepartion of drug formulations
- Those that yield chemical compounds of known structure, in themselves of no value as drugs, but useful as pharmacological tools in that they contribute to a better understanding of the mechanism of action of other drugs
- Those that have a widespread use as "herbal remedies" ("teas" and the like), in which case the medicinal value may well not have been established

In 1974 data were obtained from a National Prescription Audit (Farnsworth and Morris, 1976) as to the frequency of use of all drugs (synthetic and natural) dispensed from community pharmacies in the United States during the period 1959-1973. It turned out that prescriptions containing plant-derived drugs accounted for about 25 percent of all prescriptions and there is every indication that those containing plant-derived active constituents will remain at about that level for many years. In 1974 the retail value of prescriptions containing plant-derived active ingredients was about $3 billion. There is also abundant evidence that the "herbal tea" market, now at the multimillion-dollar level, will continue to expand rapidly.

If one groups the important drug plants--"important" in this context having reference to therapeutic effect-- three categories emerge: (1) those of major importance as prescription drugs or as sources of prescription drugs (e.g., steroids, codeine, atropine, reserpine, quinine); (2) those of somewhat lesser importance that yield compounds used as prescription drugs (e.g., colchicine, papain, castor oil, cocaine); and (3) plants yielding materials used in pharmaceuticals in various ways (e.g., gum tragacanth, gum acacia, licorice, vanilla, peppermint oil, anise, strychnine).

It could well be argued that preservation of drug plant diversity is not only important as a reservoir for new and as yet unrecognized compounds, but that the relatively recent recognition of psychotherapeutic drugs lends a special urgency to this issue. In much the same vein, it appears that an intensified last look, so to speak, at folk medicines should be undertaken before these primitive societies wholly disappear, lest some effective remedies already identified in their cultures be needlessly overlooked. And it might be well to establish a screening program and inventory of herbarium specimens for medically active compounds, with special emphasis on ecotypes of areas that lie in the path of urbanization or agricultural development.

Not all materials of pharmaceutical value come from terrestrial flora, of course, although this is the traditional source. A number of highly active chemical compounds have been found to occur in marine species of plants and animals. They include antibacterial, antiviral, and tumor-inhibiting substances, some anticoagulants, and neurobiologically active materials. Many of these substances occur in tropical species, reflecting the fact that highly diverse ecosystems characteristically provide the greatest potential source of new compounds. Yet the biota of tropical coral reefs and the diverse fauna of the deep sea is almost totally unexplored in this regard.

LIVESTOCK GERMPLASM

The issue of livestock germplasm preservation has been
thoroughly presented by Jewell, 1971; Bowman, 1974;
Mason, 1974; Lauvergne, 1975; Rendel, 1975; and Bereskin,
1976. A variety of agricultural, scientific, and cultural
justifications can be cited, but the principal ones bear
on the potential for increasing the efficiency of food
production.

Preservation of rare breeds assures a reservoir of
genetic variability, unique genetic and physiological
traits, and unknown genetic factors that provide genetic
flexibility in meeting new demands and forming new breeds.
Rare native breeds are often highly adaptable to special
environments, including unfavorable environments where
livestock production, in times of food surplus, has not
previously been implemented. Rare breeds may be useful
in crossbreeding to produce maximum hybrid vigor, to
overcome lack of adaptability in highly productive breeds,
or to meet changing product demands.

The question of preserving germplasm resources for
livestock has received more attention outside of the United
States than within. FAO reports of 1967, 1969, 1971, and
1973 considered many aspects of the problem and included
some recommendations for improvement. Preservation is a
larger challenge in older countries where native local
breeds have been developed and selected, often over centuries, to meet special needs or to fit particular environments. Many of these breeds might be useful in increasing
the efficiency of livestock production in the United States
if ways could be found to import them without danger of
introducing exotic diseases. Changing husbandry practices,
such as the shift from species fed largely on grain to
those fed on nongrain feedstuffs, increases the need for
the importation of exotic germplasm.

The importation of livestock germplasm free of exotic
diseases makes it possible to obtain unique genetic characteristics not present in domestic breeds, such as high
fertility and milk production in sheep, high growth rate
in goats, and high productivity in rabbits and other
species. Breeds with high production efficiency but
adapted to special environments in other parts of the world
could be used in areas in the United States with similar
environments. Gains from crossbreeding may well be en-
hanced by using exotic, highly productive breeds that differ genetically from domestic breeds. Preservation of
germplasm in the country of origin, until the special

characteristics are more completely measured and described, would be advantageous.

Mason (1974) has listed criteria that determine what breeds should be preserved, such as indigenous breeds, local productive breeds, genetically unique breeds, bizarre or beautiful breeds, and historically important breeds. Research needed for more adequate preservation of livestock germplasm includes cryopreservation of sperm, ova, and embryos and studies on superovulation, media, freezing, handling, storage, disease control, artificial insemination, and transplanting of embryos.

The preservation of livestock germplasm is in need of substantial support. The formation of a trust to preserve rare breeds in the United Kingdom has been described by Bowman (1974); it may be the most desirable solution to the problem, particularly if a permanent endowment with some involvement of public agencies can be established.

Breeds and strains in danger of being lost in the United States are considered more fully below.

Sheep

Three strains are in danger of being lost in the United States: Karakul, Old Type Navajo, and Southern Native. The Karakul is the only fat-tailed, fur breed of sheep in the United States. The Old Type Navajo was the only truly coarse-wool U.S. type that was adapted to very rigorous conditions; it appears to have been lost already. The Southern Native is unique in its adaptation to subtropical conditions and for its tolerance to internal parasites. It has been argued that the Southern Native might be saved if invasion by the coyote is halted.

Swine

At present there is nowhere a coordinated effort to pre-serve germplasm from breeds or strains of swine. Such an effort is needed, with initial emphasis on deciding which breeds or strains merit preservation. In addition, more research is needed to determine current production levels as a standard against which to measure yields in the future. An increase in the reproductive rates and a decrease in the generation time would reduce production costs.

Cattle

At present there seems to be no special concern about loss
of strains. Still, research is needed to better evaluate
strains and to preserve frozen germplasm.

Other Mammals

Preservation of the goat and rabbit probably should receive
attention, although hard data are lacking as to need. It
is assumed that preservation of the horse is assured by
private industry, but this may not be true of work horses.
Other farm animals, such as domestic dogs and cats, other
pets, and fur animals have not been considered.

Poultry

The only currently successful method of maintaining germplasm stocks for birds is by maintaining breeding colonies.
For chickens, turkeys, and a large number of other domesticated birds, this is accomplished by numerous fanciers
throughout the country. Some commercial breeders retain
up to 100 lines that they think might be useful in the
future. Some universities maintain a few lines, usually
not more than a half dozen or so, for their own work. The
U.S. Department of the Interior is involved in propagating
certain endangered species. There are now no germplasm
stocks, and no plans for such stocks, in any federal agency.
    Poultry producers badly need to use artificial insem-
ination because, as a result of selection, some turkeys
and meat-type chickens have such large breasts and are so
heavy that they are unable to breed naturally. Natural
mating results in extremely low fertility in turkeys.

Bees

There are probably 20,000 species of bees, worldwide.
Most of these pollinate one or more species of flowering
plant and are important as a source of honey.
    In 1922 a law was passed in the United States that pro-
hibited importation of adult bees except from Canada. It
was designed to prevent the introduction of a mite that
had been harmful to bees in Europe. Importation for research has been allowed occasionally.
    There is wide genetic diversity among honey bees.
Several years ago, a stock center was developed at Baton

38

Rouge in conjunction with the Bee Breeding Laboratory
operated by Roberts and Macheson (ARS) and an effort made
to accumulate representative stocks of bees. Inbred lines
are maintained, and queens are sold from these for breeding or research purposes. Maintaining inbred lines is
difficult because egg viability decreases rapidly with
inbreeding.

There has been some success in maintaining and shipping
semen, which should expand the gene pool.

PESTS AND PATHOGENS

Extensive research and development has been devoted to the
eradication or control of numerous pests and pathogens.
An understanding of the role that these species play in
the balance of nature is of utmost importance.

At least some of the effort expended for developing
pesticides and other control chemicals has been misdirected.
Most pesticides adversely affect members of natural communities, including humans. It is becoming increasingly
clear that more effective control of pests and pathogens
must be sought through an understanding of their biology.
This of course includes their life cycles, physiology,
reproduction, population dynamics, and ecology.

The status of knowledge about pests and pathogens is
still superficial, despite the volumes of research literature that have been produced. It appears that years of
additional effort must be invested to achieve effective
management and control of injurious species. A comprehensive collection of pests and pathogens would contribute
significantly to research on biological control.

MARINE FISH

Fishes remain among the least known groups of vertebrates,
despite their potential as a food source and the important
roles they play in aquatic ecosystems (Mayr et al., 1974).
A large task lies ahead in completing the inventory of the
fishes of the world. Many undescribed fishes are marine,
but inadequately known continental areas, such as the
Amazon Basin, doubtless have many unknown species.

Fisheries experts have documented many instances in
which races or entire species of commercially important
fishes have become extinct in a locality. In most of
these instances, extinction can be traced to the activities of men. Usually, overfishing is not the sole cause;
the situation is exacerbated by a combination of changes

in the physical or biological environment that leads to
species decline.

The complexity of the situation may be illustrated by
the Atlantic salmon. One species is found throughout
Europe and North America, but it is estimated to comprise
at least 500 distinct stocks, many of which are found in
a single tributary of a river system. The strong homing
instinct in this species ensures that adults will return
to their stream of origin, a behavior that contributes to
the formation of distinct breeding stocks. Certain stocks
have already been lost (e.g., in the Connecticut River),
and difficulties have been encountered in restocking with
fish adapted to other rivers. The situation is made worse
by mass rearing of salmon in hatcheries, which in turn
reduces the genetic base.

Genetic research on the Atlantic salmon and on other
important fish species is of relatively recent origin.
At the Atlantic Salmon Research Station (New Brunswick),
directed by Dr. Richard L. Saunders and funded by Fisheries
Marine Service, the Department of the Environment, Canada,
a program has been established to examine the fine-scale
variation in breeding structure and the range of genetic
variation in the Atlantic salmon. The life history and
migration patterns of stocks from different streams will
be characterized. One aim is to develop a relatively
opportunistic strain that can be used to colonize a variety
of rivers from which unique stocks have already been
eliminated.

Similar programs could be cited for other fish species
with which scientists are working in an effort to recoup
the loss of strains that were well adapted to given habitats. In all cases it appears that the species or stocks
that can be maintained in hatchery breeding programs
represent only a tiny fraction of the total genetic variability available in nature and that this variability can
be preserved effectively only by preserving the natural
habitats of as many species as possible. For certain
species it may be found worthwhile to go to heroic lengths
to preserve some portion of the germplasm in frozen banks
of eggs or sperm, but this will require an intensification
of research into criteria for cryopreservation of non-
mammalian gametes.

MARINE INVERTEBRATES

The marine environment supplies not only fish but also
lobsters, shrimp, crabs, clams, scallops, oysters, quahogs,

and many other commercially important species. The list
is long and undoubtedly could be extended by the exploitation of hitherto little-used species. The United States
lags behind most other highly developed countries of the
world in its use of aquacultural techniques for the improved management of these species, and the Committee on
Aquaculture, Board on Agriculture and Renewable Resources,
Commission on Natural Resources, National Research Council,
is preparing a report that will list the reasons why this
is so. A case study of the failure of many aquacultural
projects in Florida (F. T. Mannheim, personal communication) indicates that the reasons are numerous, including
certain legal barriers, the effects of vandalism, and
environmental perturbations. Not the least, from the
scientific point of view, is a lack of fundamental knowledge of the biology of the exploited species.

The most successful aquacultural projects have involved
oysters and salmon, species whose biology has been comparatively well understood for more than 50 years. Nevertheless, culture of other marine species is being attempted,
often without any fundamental understanding of the extremely high levels of genetic variability that characterize these species in their natural setting, and without
an understanding of the consequences of inbreeding and
selection under hatchery conditions.

It seems likely that few of the stocks of marine species
that have been developed and maintained by individual
workers would qualify for sustained long-term federal
support (for example, from the National Science Foundation).
Indeed, there are so few workers in marine genetics that
most species have been studied by only a few investigators, some species by only one. Such stocks, no matter
how faithfully they are maintained, do not receive sufficiently intensive use to justify federal funds. Yet,
given the paucity of genetic information on marine species,
it may be wise to develop a mechanism for ensuring some
modest long-term support of genetic studies on marine
species. Many of the food organisms used in aquacultural
projects--for example, diatoms, *Artemia salina* (brine
shrimp), and *Brachionus plicatilis* (rotifer)--have short
generation times and would lend themselves well to genetic
studies. Species of marine bivalves (e.g., oysters) have
been maintained under hatchery conditions through many
generations, but the genetic component in such studies
has often been weak, or the stocks have had to be discarded for lack of sustained financial support.

5 PRESERVATION OF ECONOMICALLY IMPORTANT MICROORGANISMS

BASIC CONCERNS

When we considered the status of conservation of microbial germplasm, we were struck by two major concerns: lack of continuity of preservation and lack of ready availability of stocks. Both stem from organizational and managerial shortcomings in relation to existing collections of cultures. There is little reason to feel that current resources are grossly inadequate for present needs, but some gaps in coverage seem to exist, and others will certainly appear if existing resources are not conserved.

Lack of Continuity

Collections of great importance are at risk because there is inadequate organization at the national level. Such losses leave us in a vulnerable position in certain areas and are extremely wasteful.

Lack of organization incurs lack of continuity in the maintenance of collections. When no continuity is provided, collections suffer to an unnecessary extent from such things as changes in supervision, shifts in responsibilities and emphasis, death or retirement of curators, and shrinking research grants. Collections in government agencies are especially likely to deteriorate when breaks in continuity occur. These agencies assemble collections to meet a social need, or as preparation for tackling an acute problem (such as coping with a disease), or for use in conducting long-term research. When, on occasion, the collections are no longer needed for the purposes for which they were assembled, they may be discontinued. Universities seldom assume responsibility for

maintaining collections assembled by their faculties, such
that when an investigator dies or retires, there is too
often no national repository to which his material can be
sent, no agency responsible for arranging to have it continued, and no means to retrieve what is unduplicable
material, whilst discarding the "usual."

In some areas, valuable resources are held in the laboratories of dedicated research workers who are attempting
to respond to national or international needs, supporting
collections on dwindling research funds at considerable
sacrifice of their time and energy. If these resources
were identified and their importance acknowledged, more
realistic arrangments could be made. Clearly some organization at the national level is required.

Lack of Availability

Many existing collections are difficult to locate; often
major collections are not generally available. This
situation often leads to underutilization of resources and
to costly duplication or hampering of activities elsewhere.

Many of our most valuable collections are used primarily
as in-house resources. They are unpublicized, their holdings are not listed, and requests for access to them are
often not granted. This is understandable, for they are
not national, public collections, and staff and funds are
not available for handling the distribution of stocks and
providing the information that must accompany them if
they are to be useful. The result is that these resources
are unavailable to other institutions; either the holdings
must be duplicated elsewhere, or research and development
is hampered. The discarding of a collection in one
laboratory is an intolerable waste if another laboratory
is meanwhile busying itself assembling a similar collection.

In the past resources were ususally located by word of
mouth or through notations appearing in published scientific papers. These procedures were satisfactory as long
as the known organisms and the people working with them
were few, but within the past 50 years these devices
ceased to be adequate. No one can hope to keep track of
more than a small fraction of the work published on microorganisms or to be aware of the location of more than a
small fraction of the collections. Work with microorganisms is now conducted on such a scale, involves so many
people, and affects society in so many ways that its
management calls for improved organization. At present

each large laboratory generally attempts to assemble its own collection and to be self-sufficient; the small laboratory is hard-put to find adequate resources to do this.

The question therefore is: How can we set up a more effective system to replace the traditional one, which broke down years ago?

One possibility is to create a single mammoth public collection of cultures or microorganisms for the United States. This was the solution envisaged 50 years ago when the American Type Culture Collection (ATCC) was founded. The organisms deemed worthy of preservation then numbered a few hundred specimens. We now know that microorganisms are extremely numerous, are extremely diverse (more so than any other group of organisms), and benefit society in many previously unsuspected ways. It is very likely that hundreds of thousands of specimens are worthy of preservation.

Is the approach exemplified by ATCC practical today? The most telling argument against it is that it probably would be impossible to assemble in one spot the curatorial expertise needed to handle the variety of organisms and the types of data that would have to be handled. It seems much more reasonable to build upon the existing scattered facilities as much as possible, seeking some consolidation and some overall organization to ensure long-term continuity and adequacy of coverage, availability, and quality.

PATHOGENS

The germplasm resources essential to the control of infectious diseases include large, comprehensive laboratory collections of pathogenic microorganisms. These organisms cannot always be obtained at will from nature and they are constantly needed in diagnosis of disease, for development of therapy and control measures, for selection or development of resistant host strains, for the study of epidemiology, and for the study of fundamental pathogenic mechanisms.

Because the task of preserving pathogenic microorganisms is worldwide, it can best be approached by organization at national and international levels. Such organization is needed to ensure adequate coverage and safety and to avoid excess duplication. The system must be sufficiently flexible to cope with changing social conditions, the movement of people, the introduction of new species, and the evolution of hosts and pathogens.

It is essential that the major collections be situated in laboratories having stable support and continuity of expert supervision. The need to preserve the agents responsible for past outbreaks of disease, including all significant variants encountered, makes for steady growth of these collections. Great care must be taken in deciding whether and when to reduce the holdings or to transfer their supervision, although constant culling is necessary, as in all collections.

The World Health Organization (WHO) has assumed responsibility for organization, at the international level, of germplasm resources essential for the control of infectious diseases of man. WHO has designated certain laboratories around the world as international and regional reference centers and has charged them with responsibility for maintaining reference collections of specific groups of microorganisms for the entire world (WHO, 1969).

About one-fifth of the centers described in 1969 were in the United States--in government laboratories, universities, and private institutions. To some extent the centers are supported by the institutions in which they are located and by grants from governmental and private agencies. Funds available from WHO are insufficient to support them.

In the United States a chaotic situation results from the fact that there are no national centers charged with responsibility for maintaining comprehensive culture collections of pathogens for use in the control and study of disease in man, plants, and animals. There are, indeed, large, comprehensive collections that serve these needs, but they are widely scattered; it is difficult to learn what their holdings are; and furthermore the holdings are subject to frequent and drastic changes. The collections have been built up in laboratories of the U.S. Public Health Service, the U.S. Department of the Army, the U.S. Department of Agriculture, and other government agencies; in university and other research laboratories; and in commercial laboratories.

MICROORGANISMS OTHER THAN PATHOGENS

Economically important, nonpathogenic microorganisms include:

• Symbionts of plants and animals of economic importance, which include some of the nitrogen-fixing species

- Agents used in food processing, as in the dairy and brewing industries
- Organisms that serve as direct food sources for man or domestic animals
- Organisms that may be used in controlling insects or other pests
- Organisms that are involved in food spoilage, or destruction of fabrics, fibers, forest products, and other materiel
- Organisms employed to process wastes (e.g., sewage purification, treatment of industrial plant effluents, conversion of wastes into useful products)
- Organisms used in producing chemicals (e.g., pharmaceutical products such as antibiotics and steroids, and industrial organic chemicals)
- Organisms used in monitoring the environment for the presence of harmful chemicals, such as mutagens
- Microorganisms used in applied and basic research

The germplasm resources on which man depends in dealing with nonpathogenic microorganisms consist primarily of collections of cultures in laboratories. These collections are a valuable investment. Many of the cultures are strains of microorganisms that have been selected and developed for specific economically useful properties, and therefore correspond to the crop plants and domestic animals among the higher organisms.

The loss of the working collections of microorganisms on which public health services, agriculture, and industry depend would lead to considerable disruption of these activities. Their replacement from nature would require years of effort and very large expenditures of funds. Thus it is essential to take such steps as will efficiently preserve what we have.

An evaluation of the major collections of cultures of microorganisms in the United States would be difficult, but it should be performed. Earlier efforts to compile lists of these resources failed to achieve complete coverage and did not assess the quality and value of the collections listed. It is impossible to describe the situation accurately on the basis of available information, but what is known suggests waste, inefficiency, considerable duplication in some areas, and perilously sparse coverage in other areas.

The ATCC, set up in 1925, has been noted above as the one widely recognized culture collection in the United States. It is a private, nonprofit organization supported largely by government agencies, by fees charged to recipients of stocks, and by fees charged to donors for maintenance or distribution of stocks. Although it contains a wide variety of useful microorganisms, it probably could not serve as the sole national resource for any group of microorganisms.

Some industrial collections are among the largest and most stable. A few have more extensive holdings, numerically, than the ATCC. An idea of the extent to which our society is dependent on industrial collections can be gained from a review of the fermentation industries producing pharmaceuticals and fine chemicals, worldwide (Perlman, 1977). This review lists 183 products, including 73 antibiotics, 18 tetracyclines, 27 enzymes, and 10 organic acids, plus solvents, vitamins, amino acids, and steroids that are produced by microbial fermentation. Some of these pharmaceuticals and chemicals are produced in quantities of 50,000-200,000 tons per year. Furthermore there is increasing interest in using microbial cells for protein in human and animal diets and in using microbial processes to generate chemicals now produced chemically from petroleum products.

The remaining microbial collections on which the nation depends are scattered in laboratories in government agencies, universities, and elsewhere. The questions of continuity and availability are similar to those afflicting collections of pathogens. These collections are difficult to locate, which in turn leads to underutilization and duplication of effort. Continuity often depends on the efforts of dedicated persons who accept responsibility for assembling and maintaining the collections and for obtaining financial support. Usually the collections are supported by funds that have not been specifically designated for that purpose. These collections, like those of pathogens, sometimes are discarded upon the retirement or death of the caretaker.

Certain factors contributing to this unsatisfactory situation are: (1) Lack of organization at the national level and consequent lack of support and clearly assigned agency responsibilities; (2) a general mistrust of centralization, which results in part from past instances of uncertain support and inadequate supervision; (3) failure to develop a system of data storage and retrieval adequate for the laboratories, activities, and data now existing; (4) an economic situation of some years' duration that provided research funds at a level such that extensive, but uncoordinated, culture collections could be built up

and maintained on research grants not specifically designated for that purpose.

The prospect of greater dependence on microorganisms
for food and chemicals lends urgency to the need to assess
our resources and to provide for their use and preservation
in a more efficient manner. The problem is exacerbated by
trends in science education that have deemphasized broad
biological training required for those reponsible for maintaining collections of broad coverage.

Some efforts to rectify the situation have been made.
Internationally a start has been made in compiling a list
of collections of cultures of microorganisms. A World
Federation for Culture Collections (WFCC; formerly, Section on Culture Collections) has been formed within the
International Association of Microbiological Societies of
the United Nations Educational, Scientific, and Cultural
Organization (UNESCO). WFCC sponsored a worldwide survey
of culture collections with support from UNESCO, the World
Health Organization, and the Commonwealth Scientific and
Industrial Research Organization, Australia. A world
directory was published as a result of this survey (Martin
and Skerman, 1972). The 349 collections listed in the
directory represent only a small fraction of the existing
collections, but they are a beginning. Efforts are being
made to establish an international information center to
provide information on collections (see Skerman, p. 141,
in Iizuka and Hasegawa, 1970). Coverage for the United
States, however, is inadequate; many important collections
are omitted and some trivial ones are included.

The difficulties encountered in conducting the survey
indicate how great was the need (Martin and Quadling, 1970,
p. 133). Even the microbiological societies of the various
countries apparently were of little help in this respect.
Those in charge of many collections did not respond to
questionnaires, in part, perhaps from apprehension that
to be listed in the directory would generate more requests
for cultures than could be handled (Simmons, 1970).

UNESCO has also supported international conferences
on culture collections. The first was held in Tokyo in
1968. The proceedings (Iizuka and Hasegawa, 1970) suggest
that the United States is not meeting its needs for microbial culture collections as well as are a number of other
countries (e.g., Japan and the United Kingdom).

The WFCC has placed considerable emphasis on the development of culture collections and the expertise necessary
for managing them in developing countries.

6 COLLECTIONS FOR RESEARCH, TEACHING, AND PUBLIC EDUCATION

The status of collections of noneconomic significance varies widely, depending on the type of organisms involved and the use to which each collection is put. The size of a collection, the form in which it is maintained, its value in terms of the effort that went into its development, its future usefulness, and so on, are factors that must be considered. We must, however, recognize the value of collections in a general sense and the need to ensure their high quality and continuity.

Research collections of organisms used as models for exploring the fundamental properties of life are an irreplaceable resource because of the mutant strains that have been accumulated and the knowledge that has been developed in the course of their characterization. Because of these background data such strains can be used in the investigation of problems whose solutions would be of signal benefit.

Less widely used and more specialized collections, particularly those developed by a single investigator or fancier, are far less secure than collections of economically important organisms because much of the value is represented by the knowledge and skill accumulated by that one person. To recognize the value of such a collection is at times difficult; even when value is obvious, it is often difficult to find someone else able and willing to carry on the work.

Research collections, in addition to requiring specialized knowledge, raise a variety of problems as to quality control and maintenance. Some require elaborate conditions for culturing and preservation.

48

BOTANICAL GARDENS AND ARBORETA

The role of botanical gardens and arboreta as institutions of fundamental importance in the conservation of genetic resources is too frequently overlooked. Technically speaking, arboreta grow only woody shrubs and trees, whereas botanical gardens are not thus restricted; in practice, the differences are trivial.

Most who are familiar with the important botanical gardens and arboreta have been struck by the uniformity of their traditions and functions. In the Americas and Europe, if not elsewhere, the gardens are largely nineteenth-century institutions created to meet the demand for popular education in a period when world travel was restricted to the few. They parallel Victorian museums; in many respects museums and gardens evolved together.

Botanical gardens early became citadels of research on diversity and the cultivation and propagation of plants. They were not thought of as instruments of plant conservation, mainly because at the height of their expansion Victorian optimism extended over the whole field of world resources, which then seemed inexhaustible. Although rare plants were often cultivated, there was little concern for preservation of the stock, and records were hardly ever maintained. Unfortunately, most gardens continue to operate on this level; such conservation role as they fill is ususally secondary and is seldom well documented. An increasing number of botanists deplore this state of affairs, and their swelling ranks now include some who direct botanical gardens and determine their program priorities.

What in fact can botanical gardens and arboreta do to promote a better understanding of the endangered-species issue and actually to conserve genetic resources? By their very nature they are treasuries of botanical diversity, growing everything from algae to redwoods and displaying a wide array of native and exotic species and cultivars. It is not uncommon for them to maintain several thousand species in a score of natural or simulated habitats. Regardless of the quality and extent of documentation, the mere fact that such an aggregation of genotypes is consistently maintained gives it conservation value and argues for a more direct role in botanical conservation. Several gardens also own, or have custody of, significant tracts of natural vegetation that promote public understanding, serve as refuges for the genetic resources found in them, and make these resources available for research.

Historically, botanical gardens have often been very

active in distributing plant materials. Most have an
Index Seminum and have participated in international exchanges on a considerable scale.

Associated with the living collections of many botanical
gardens are libraries devoted to botany and horticulture.
Several institutions also maintain plant-specimen museums,
or herbariums, which are collections of well-documented
specimens representing a very diverse flora. Thus gardens
and arboreta and their associated facilities are essential
instruments in research in plant systematics, evolution,
geography, and ecology. Such research is fundamental not
only for understanding the forces that accelerate impover-
ishment of genetic resources, but also for establishing
sound conservation practices.

Botanical gardens customarily offer educational programs to an increasingly interested lay public. Several,
either as components of or in association with universities,
provide professional and postgraduate training in botanical
science and in various aspects of horticulture. Both
activities provide a sound basis for developing a stronger
conservation focus.

Public Exhibits

Gardens and arboreta should develop public exhibits drama-
tizing the need for botanical conservation. In this respect, the custodians of the plant world have much to do.
Few are aware that one in 20 native plant species in
North America is threatened with extinction. Displaying
and interpreting rare plants could give enormous publicity
to local, national, and international conservation organizations. By responding to such questions as why some
plants are rare, what factors threaten their continued
existence, and why botanical diversity is important, gardens could do much to help organizations involved with the
preservation of natural resources. No other group is
better able to espouse the cause of botanical diversity.

Research Emphasis

Botanical gardens and arboreta should broaden or shift
their research programs to include studies on plant ontog-
eny, longevity, and reproductive effectiveness, focusing
these investigations on rare plants. Information of this
kind is essential to any attempt at salvaging plant species

threatened with extinction by bringing them into cultivation. So little information of this nature is available
that attempted salvage often fails. Simple, well-documented
cultivation experiments, preferably in conjunction with a
nature preserve, would do much to provide information on
species of narrow ecological amplitude whose survival is
dependent on cultivation in alien habitats.

Preservation of Endangered Species

Botanical gardens and arboreta should do a better job of
safeguarding stocks of endangered plant species. Critical
to this function is a satisfactory system of documentation
for living collections. Wild species in cultivation, and
all seeds and other propagating materials sent out on exchange to other gardens, should be of known and recorded
origin. Propagation of rare and endangered species should
be attempted not by transplanting (unless this is the only
remaining option), but by judicious seed or cutting collection. American botanical gardens and arboreta are the
most logical agencies to take on the responsibility of
protecting the thousand-odd plant species proposed for
threatened and endangered status in the United States.
Special attention should be given to species actually
facing extinction, to narrowly endemic species confined
to a particular limited area, and to wild relatives of
important cultivars. Some species may lend themselves to
seed-bank conservation; others must be maintained in successive generations. Either approach should be undertaken
with a view to eventual reintroduction to the wild, where
possible.
    Because some larger gardens have long-standing involve-
ment with programs in the tropics, where both speciation
generally and the difficulty of protecting rare and endangered species are greater (in part because so little
is known about them), they might well emphasize the
preservation of selected tropical groups. International
coordination is obviously essential to success in such a
complex endeavor, and several conferences addressing this
topic have been held in recent years.
    A number of questions confront botanical gardens and
arboreta with respect to threatened and endangered species:

• How can arboreta and gardens serve as salvage centers for species otherwise certain to become extinct?
• How can the already slender and diminishing financial

resources available to the gardens and arboreta be directed
to such a purpose?
• How can the salvaging of threatened and endangered
species serve more than an archival function?
• How can arboreta and gardens assist other agencies
in preparing for the reintroduction of salvaged species
into the wild?
• How can arboreta and gardens function in enlarging
the catalog of economic plants?

Botanical gardens face new challenges and problems that
cannot be met in the old way. The new spirit requires
integrated, coordinated efforts by the world community of
botanical gardens and arboreta. That approach is given
impetus by the increasingly prominent roles played by the
United Nations (as in the Man and the Biosphere program)
and by the International Union for the Conservation of
Nature and Natural Resources (IUCN), the World Wildlife
Fund, the Nature Conservancy, the National Audubon Society,
the International Association of Botanic Gardens, and
others.

ZOOS AND AQUARIA

An increasing number of wild species face extinction. The
reasons for this threat are detailed by Zisweiler (1967)
and Myers (1977), who provide cogent arguments why the
loss of species is of serious concern.
Zoos and aquaria now safeguard some of these species;
others can find a refuge if sufficiently coordinated efforts are made. But for many animal groups it is impossible thus to avoid their extinction. Extinction of whales
and certain other marine life can be prevented only through
international agreements. Some migrating species of birds,
e.g., hummingbirds and the whooping crane, can be conserved only by preserving their very limited habitats.
Ecological niches are so specific for some species, and
their reproduction in captivity is so complex or poorly
understood (e.g., bats and the platypus), that preservation of wild habitats is the only means of ensuring preservation. The national refuge system, the national park
system, and various other schemes seek to meet the need,
but in many cases the ecosystem harboring a species is
already destroyed or unsafe.

Germplasm Resources

Zoos and aquaria represent a substantial industry; the
United States alone has 230 listed in the *International
Zoo Yearbook* (Olney, 1976). Together, they hold many
species that have become extinct in nature (best known are
Pere David's deer, Arabian oryx, and Przewalski's horse)
and others that are severely threatened.

Intensive efforts have accomplished reintroduction of
certain species to their original habitat. The Hawaiian
goose (*Branta sandivicensis*) is a good example (Martin,
1975). In 1700 there were about 25,000, in 1940 only 43.
Today, as a result of captive breeding, there are over
3,000 in the wild and 200 in captivity.

Traditionally, zoos have replaced losses by purchase
from dealers, but in the last decade serious efforts at
conservation have been started, and reproduction of
threatened species in captivity is of primary concern to
modern zookeepers. The International Species Identification System (ISIS), a computerized program located at
Minnesota State Zoo, gathers data on animals held in zoos
and aquaria for breeding purposes. Over 18,000 are now
recorded, from 111 participants. Two international con-
gresses on breeding endangered species in captivity have
been held (Martin, 1975; Olney, 1977).

The accelerating rate of extermination is correlated
with the increase in human population. Zisweiler (1967)
names the following specific causes: direct extermination
from hunting for meat, eggs, hides, furs, and feathers,
souvenirs, superstition, trade or research, trophies; in-
direct extermination by destruction of natural vegetation,
drainage of wetlands, ravaged water, air and water pollution, radiation, traffic, domestic animal disease, and
changed flora and fauna; introduction of competing animals.

In light of current estimates of future human expansion
and the destruction of tropical rain forest, it can safely
be anticipated that many species will vanish in the next
decade unless their habitats are conserved. In the past
two centuries at least 101 bird and 46 mammal species have
become extinct because of human intervention (Zisweiler,
1967). The Red Data Books (IUCN) provide an extensive
analysis of mammals, reptiles, amphibians, and fishes in
various degrees of endangerment, and a comprehensive review on amphibians and reptiles has been compiled by
R. E. Ashton (Edwards and Pisani, 1976). The Fish and
Wildlife Service, U.S. Department of the Interior, issues
a monthly bulletin whose purpose is to keep the public

54

informed of the rapidly changing situation. The following
listing is taken from the February 1977 issue of the bulletin (Fish and Wildlife Service, 1977):

Category	U.S.	Foreign	Total
NUMBER OF ENDANGERED SPECIES			
Mammals	36	227	263
Birds	66	144	210
Reptiles	8	46	54
Amphibians	4	9	13
Fishes	30	10	40
Snails	0	1	1
Clams	22	2	24
Insects	6	0	6
Total	172	439	611
NUMBER OF THREATENED SPECIES			
Mammals	2	17	19
Birds	1	0	1
Reptiles	1	0	1
Amphibians	1	0	1
Fishes	4	0	4
Insects	2	0	2
Total	11	17	28

The reproduction of vanishing species has become a self-
imposed duty in most major zoos and aquariums. There are
formidable problems. Only a small proportion of certain
species will breed in captivity. Thus there is very little
whale reproduction in captivity, although it is feasible
(Ridgway and Benirschke, 1977). Virtually no bats are
bred in zoos, and only about 10 percent of captive reptiles
propagate. Hummingbirds and penguins are notoriously difficult to breed.

For many endangered or threatened species, captive propagation is not feasible. The space and cost requirements
for caging and care would be astronomical. Consequently,
zoos and aquaria have accepted the challenge to propagate
a selected number of severely threatened species that are
also useful for exhibition. For many of these species,
some basic biological information is needed (e.g., convenient techniques to determine sex in birds) (Olney, 1977).
For some species (e.g., Przewalski's horse), genetic
information is needed to prevent inbreeding depression.
Others (e.g., Siberian tiger) are subject to diseases in
captivity that need to be studied and eliminated.

Many aspects of reproductive physiology and communicable diseases present technical barriers to reproduction. Zoos and aquaria are beginning to address many of these questions, and they need encouragement and support. Above all, restrictive legislation concerning endangered species and the transportation of specimens must be developed in consultation with these institutions; it must not become a further impediment to achieving reproduction in captivity.

Research

Many biomedical questions can hardly be addressed unless animal models are available. An understanding of some of the basic aspects of nature has come from the comparative study of animals--their genetics, diseases, and so on. Man has greatly profited from this study, particularly in medicine. There would have been little progress in virology and preventive vaccination against poliomyelitis without the availability of nonhuman primates. Other investigations of disease (e.g., slow virus brain diseases and tumor virus research) depend heavily on primates.

Much of this research is carried out in primate centers, but these centers suffer from drastic reduction in availability of animals from the wild. India reduced exportation in 1975 from 50,000 to 30,000 annually, South American countries have severely restricted exportation (ILAR, 1975), and Thailand has banned all primate exportation (Fitter, 1977). Certain species are unavailable for most research purposes, because of their endangered status or because supplies are unobtainable; examples are apes, golden marmosets and lemurs (Bridgewater, 1972). Hubbs and Bleby (1976), reporting to the Medical Research Council in England, find that "the laboratory primate is likely to be of rapidly growing importance in many fields of biomedical research." They continue:

Thus, even if it were feasible on health grounds to go on using wild animals, their supply would decrease with this increasing demand. The main solution to the problems would appear to be the early commencement of breeding primates along the same lines as all other commonly used laboratory species. This solution is recommended by the World Health Organization.

Progressive zoos are doing just that for all species available to them. Although few have the financial resources to undertake this massive job, staff of most zoos are deeply concerned about conservation.

Aside from primates, biomedicine gains enormously from the models found by chance in zoos and aquaria. In the comparative genetics of mammals, e.g., the lowest chromosome number (6) is found in a deer, *Muntiacus muntjac*, of which perhaps only 10 exist in zoos at present (Benirschke, 1977). This species' nearest relative has a chromosome number of 46; thus it would be very useful for genetic studies.

Funds have not been generally available for biomedical research in zoos because many institutions lacked qualified investigators and facilities, because much of the proposed research was unacceptable to zoo officials on the argument that their specimens were in the endangered category, and because of public pressures. This situation is gradually changing (ILAR, 1975), but a vast amount of material goes unused. Cell strains for mutagenesis, virus, and research on aging could be obtained from these collections. Virology on exotic species is important for human medicine and agriculture but is totally unexploited. Reproductive biology could be studied without threat to the species.

It would be possible to build up a systematic collection of cells of all varieties (see also p. 77). The cells should be conserved for use in large cell banks. The same can be said for spermatozoa, serum, blood cells, and embryos. A vast resource is not being used, and in many instances the material will not be available in the future for study by more advanced techniques. Only one strain of blue whale cells now exists, and only a total 250 fibroblast strains from other zoo mammals. There is no funding for a systematic effort to collect materials from any zoo species. Yet we can hope that through proper preservation of the nuclear genome the genetic content can be studied at a later date when extinction has caught up with the species and when nuclear transfer may become feasible. Although these prospects were highly speculative at first, it has now been shown in mice that individual foreign cell lines can be introduced into the embryo to render it chimeric and that it will reproduce this genome in future generations. Thus we can anticipate that the collection of somatic cells from vanishing species would be useful for the re-creation of these species in the future. Systematic collection should be undertaken at this time.

There is also a great need for research into reproductive
physiology, exsemination (Seager et al., 1975), ovum
transfer, and the like, and this research must be conducted in zoos and aquaria.

While primate centers concern themselves with the most
"biomedically useful" species, many other threatened species of nonhuman primates "fall between the cracks" (e.g.,
langurs, pigmy chimpanzees, lemurs, and orangutans). This
situation must not be allowed to continue.

Properly trained biologists are needed to participate
in research on captive breeding and disease control.
Public funds for such research are virtually unavilable
unless the programs have a biomedical connotation. It is
essential that the public, legislators, and federal funding agencies be educated as to the dire need for work in
this area before the trend toward extinction is further
advanced. If the importance of conservation research to
human health were adequately recognized, appropriate funds
could be channeled into the work.

There is a great need for zoos and aquaria to become
affiliated with universities and to conduct research in
reproductive biology, behavior, and disease control. What
is done for domestic and laboratory species (e.g., artificial insemination and sperm and blastocyst freezing and
implantation) should be done for all species reproducing
in captivity. Importation of animals for exhibition and
for educational purposes would be unnecessary if the
basic biological factors in reproduction were understood.
If needed research on reproduction in captivity could be
conducted, there would be less cause for concern about
the imminent extinction of such species as the Puerto
Rico parrot (whose few remaining specimens have not even
been sexed), the California condor, and the bald eagle.

Short (1976) has called attention to the possibility of
finding future domestic stock among zoo species. It is
possible that some of these species could be developed into
meat animals. Candidates include eland, beisa oryx, and
perhaps special breeds of wild sheep and pigs. Some might
prove to be desirable as meat animals because of their
food preferences, which are different from those of the
domestic animals we now have.

Regulation and Coordination

Because not all species can be artificially conserved,
choices must be made. Zoos and aquaria are prepared to

participate in this effort; these matters are repeatedly
discussed at national and international meetings and in
relevant publications. But they face restrictive legislation, severe financial problems, and often ill-advised
public pressures. Zoos are establishing priorities, and
breeding consortia are being formed, but there is no overall plan to assure that maximum conservation occurs.

Whereas new journals seek to correct the lack of awareness among scientists, only the Fauna Protection Society
and IUCN attempt to bring all available data together
systematically, but even here they lack sufficient funds
for this attempt. When, occasionally, broadly trained
biologists are available, institutions seldom have funds
to support them or the needed research. Moreover, a
flurry of new laws and regulations have severely impeded
captive breeding. While it is recognized that the promul-
gators have good intentions--and that many of the regulations have been highly effective--the regulations are so
restrictive, or the processing of forms so laborious,
that worthwhile efforts are hindered. Large zoos now
require a special staff to deal with permits as required
to effect simple breeding loans of endangered species in
our own country. Without a special permit (which is itself difficult to get), it is illegal, for instance, to
send a tissue culture of a stranded dead whale to Edinburgh,
where the only DNA research with this type of tissue is
going on.

There is little communication between government officials and the persons who do the actual breeding work or
conduct the research into reproduction of endangered species. So complex is the legislation and regulation on
these matters that the Society of Mammalogists saw fit to
publish for its members a summary of the regulations, but
warned that the report should not be used as a primary
source of information because of constant revision and
review (Genoway and Choate, 1976; ILAR News, 1976). The
rationale for and evolution of these laws is dealt with
in great detail by the Council on Environmental Quality
(Bean, 1977).

Well-intentioned captive breeding programs suffer from
overregulation, delays, and many senseless restrictions.
Better avenues are needed for communication between
regulatory offices and the centers for reproduction and
research, so that captive breeding can be vigorously
pursued. An important recent instance is the destruction
by the USDA of thousands of quarantined birds, including
even endangered species, when Newcastle disease was

identified in a single specimen. Alternative ways to deal
with this situation exist and must be implemented. Like-
wise, we have ways of ensuring freedom from hoof-and-mouth
disease and hog cholera that are unlike those currently
used.

The endangered-species program has as its primary objective restoring threatened or endangered species to
abundance. It does so primarily by regulating export and
import. A secondary objective is captive propagation in
survival centers. Conservation of germplasm is implicit.
The criteria for the selection of species under the
Endangered Species Act of 1973 are:

The Secretary shall by regulation determine whether
any species is an endangered species or a threatened
species because of any of the following factors:
1. The present or threatened destruction, modifi-
    cation, or curtailment of its habitat or range
2. Overutilization for commercial, sporting,
    scientific or educational purposes
3. Disease or predation
4. Inadequacy of existing regulatory mechanisms
5. Other natural or manmade factors affecting its
    continued existence

The actual listing is similar to that of the IUCN, although a number of exceptions exist, presumably in part
the result of pressure on the Department from special-
interest groups and of inadequate time for study. An
international treaty, implemented in 1977, severely limits
importation and exportation of wildlife.

The IUCN defines an "endangered species" as a taxon "in
danger of extinction, the survival of which is unlikely if
the causal factors now at work continue operating. These
taxa are those where numbers have been reduced to a critically low level or the extent of their habitat has been
drastically reduced so that they are deemed to be in
immediate danger of extinction." The next critical level
is "vulnerable," followed by "rare," "out of danger," and
"indeterminate." The individual species are carefully
assessed by various scientific groups and documented in
the Red Data Books, which contain information on status,
distribution, population, habitat, conservation measures
taken or proposed, and remarks.

Although the United States designations are usually
similar to those in the IUCN listing, they often differ,
and when different almost always place the species in a

more endangered category. Because the IUCN listing is
well documented, its recommendations should be followed
whenever possible by the United States, and reasons given
when deviations are considered unavoidable. The Red Data
Books of the IUCN are periodically revised, and special
groups of experts (e.g., the Rhino group) are assigned
responsibility for recommending changes.

Continuity, Security, and Technology

Zoos and aquaria need no new regulations to ensure their
continuity or to provide security for any given species.
What they need is encouragement and support. Their success
as organizations depends on their success in captive reproduction, and they are, therefore, highly motivated to pur-
sue research in this area. But they urgently need additional
scientific expertise to deal with a multitude of problems.
    There is virtually no broad-based attempt to identify
the genetic compositon of captive species, and no assurance
can be given that captive breeding will be successful and
not lead quickly to inbreeding depression. More biological investigation is needed to characterize breeding groups
and ensure meaningful exchange between gene pools. Although these questions have been addressed in two reports
(ILAR, 1975; Martin, 1975), much more work is necessary in
this area. Also, at present no specific responsibilities
for such exchange--or, indeed, for acting as guardian for
any species--are assigned to zoos or other institutions.
Breeding consortia are being formed (e.g., for the gorilla),
and the Trustees of the Arabian Oryx have assumed international responsibility for the maintenance of this species, extinct in the wild. The removal of restrictive
legislation would help these motivated groups.

ORGANISMS FOR RESEARCH IN GENETICS

Stock Collections

Breeding stocks used in genetic research constitute a
particularly important germplasm resource. Comparable
stocks used in the development of agriculturally and in-
dustrially important products are treated elsewhere in
this report, because they present different management
problems. We are concerned here primarily with the materials assembled in the course of, and for use in, basic
studies in biology and medicine.

It must be emphasized at the outset that these materials are extremely diverse, and are highly varied with respect to their usefulness and their degree of genetic definition. The most extensively studied organisms, such as *Drosophila melanogaster*, *Escherichia coli*, and *Neurospora sitophila*, are available in extensive collections containing many natural variants, selected mutants, and combinations of genes and chromosome arrangements useful for different purposes. When many investigators use the same or similar biological materials, provision is usually made for the maintenance of a stock center to serve the needs of that community and to effect savings in the storage of materials used only occasionally. Such groups of workers may also develop an information-exchange newsletter, to announce the availability of newly discovered strains, the isolation of interesting mutants, and the development of new techniques.

Many collections, however, do not represent a community effort and a shared responsibility but are the product of a single individual's research. Genetic studies on any organism begin, of course, upon the initiative of one or a few explorers. The initial efforts may lead to wider use of the organisms, but in only occasional instances does it lead to truly widespread use.

These organism-based studies are an integral and important component of our national research effort. The "domestication" of the fruit fly led to the first compelling demonstration of the role of chromosomes in heredity. Studies on *Neurospora* eventually gave the first clear insights into the genetic control of metabolism. Genetic analyses of coliform bacteria and their viruses prepared the way for the explosive developement of molecular biology after World War II. Even now explorations are under way to discover whether new organisms (algae, nematodes, or squash bugs)--or old organisms used in new ways--may open new eras in regulatory biology, developmental biology, or population analysis. These collections of more or less genetically defined materials, accumulated through much effort over many years, usually supported by public funds, are an important national resource and are the basis for continued exploration of life processes.

Exploratory efforts sometimes lead, as noted just above, to the establishment of an organism as a model system for the study of particular biological processes. But the value of the materials thus accumulated does not depend solely on their incorporation into a comprehensive model system. The study of pathologies in animal models, for example, is an increasingly important means of approaching human disease. If a genetic defect is found in a mouse or a hamster that mimics muscular dystrophy, diabetes mellitus, or cerebral palsy, the mutant strain may be of great interest in medical studies, quite aside from its value in a mouse or hamster genetic program. Although this use of genetic materials in another species to illuminate processes in man is a well-understood approach, it is but one example of the utility of the comparative approach in biology and medicine.

An instructive parallel can be drawn between the task of managing germplasm for a commercial agricultural community and that of managing the germplasm requirements of the biomedical research community. Organisms in nature are the equivalent of the wild populations of a crop plant. Collections of organisms maintained for commercial use-- in the production of cheese, wine, or antibiotics, for example--or for exhibitors, consumers, or fanciers of fish, birds, or mammals, correspond to the folk or land varieties of a crop plant. The few generally recognized model research systems constitute the payoff system of the research effort, much as commercial varieties are the culmination of the geneflow from native flora into crop plants. Between the wild populations and folk varieties and the major model systems, there are many smaller collections of organisms, more or less well defined genetically, with potentialities as major model systems. Just as genetic uniformity is an economic advantage in a commercial crop, so too it is advantageous to use a few well-chosen experimental systems in concentrated and collaborative research. And just as genetic uniformity in agriculture carries with it vulnerability to altered conditions or uses, so too can over-reliance on a few genetic models have potential dangers in research. Thus research collections of a wide variety of organisms provide the basis for new investigative directions in a vital biomedical community.

Experimental Mammals

Biomedical research depends heavily upon the availability of genetically defined animal stocks--usually, although not exclusively, experimental mammals--for increasing understanding of normal and abnormal growth, development, differentiation, function, and aging deterioration. Analyses that require planned matings or tampering with the life history can be done only under experimental

conditions. Primates, mice, and rats are, and will continue to be, extensively used in paramedical research both
because of their intrinsic suitability and because of the
"added value" of accumulated knowledge resulting from pre-
vious experimentation with the same species, inbred strains,
or mutants. For researchers to gain from the experience
of past investigators, and from special stocks that they
have created, it is necessary that a considerable variety
of genetically controlled animal stocks be permanently
maintained.

Primates   Both Old-World and New-World primates are very
important in medical research. Collection of primates
from the wild, which are then imported into the United
States, has until recently been the chief source of research primates. But the eventual need to breed all nonhuman primates for biomedical research is now accepted
(Bermant and Lindburg, 1975). For comparison with humans,
primates are especially important in studies involving the
nervous system. (See also p. 55, *et seq.*) Several different primate species, and variant populations within the
squirrel monkeys, provide excellent material for the study
of atherosclerosis. The reproductive physiology of primates corresponds more closely with that of humans than
does that of any other animal group.

Monkeys are slow to mature and have few offspring so
that establishment and maintenance of breeding colonies is
necessarily slow and expensive, but is nevertheless essential for future supply of primates for research. This
solution for research primate supply is recommended by
the World Health Organization.

Mice   Laboratory mice are very important in biomedical
research, largely because so many and diverse inbred
strains and mutants are currently available. Inbred lines
(derived from more than 20 successive brother-sister matings)
tend to be very uniform, including having the same histocompatibility genes, the same isozymic forms of enzymes,
and a tendency for the same types of pathological lesions.
These mice, plus special congenic stocks in which specific
alleles characteristically found in one inbred strain have
been transferred to a different genetic background, are
very widely used in immunogenetic and other immunological
research. All mammalian species thus far studied have one
major histocompatibility locus, similar to the human HLA,
and numerous minor loci. More is known about effects and
genetic fine structure of the major *H-2* complex locus of

64

the mouse, which corresponds to human HLA, than about any
other histocompatibility locus in any mammalian species.
The complex has effects on immune responses as well as
transplantation. Continued advances in immunology will
be greatly helped by continued maintenance and availability
of a number of special genetic stocks carrying recombinations within the *H-2* locus. Such recombinations happen
very rarely, but each must be kept as a critical research
tool. A great deal of research on T- and B-lymphocytes
and their interactions uses special congenic mouse stocks.

A predictable proportion of the mice in any one inbred
strain develop lung tumors, mammary tumors, lymphatic leukemia, Hodgkin's-like lesions, amyloidosis, plasma cell
tumors, or other specific degenerations and thus provide
valuable tools for cancer or aging research. The inciden-
ces can often be elevated by carcinogenic agents, thus
providing test animals for experimental carcinogenesis or
for screening of chemicals. They are also valuable for
screening of potential mutagens. The use of appropriately
sensitive strains, and of a variety of strains, is important for obtaining conclusive evidence.

Deleterious mutants that occur in mice, whether spontaneous or induced, provide valuable tools for study of
constitutional diseases that correspond to, or are at
least similar to, human hereditary diseases. These mutants, of which a very large number are known, are most
valuable if they occur in otherwise genetically homogeneous
inbred mice, since the availability of a single gene difference segregating against a homogeneous genetic background greatly facilitates analysis of the action of the
mutant gene. Because each mutation is a rare event, it
is important that it be recognized, preserved, characterized, and exploited for its bearing on medical problems.

Rats   Laboratory rats have been widely used in research,
especially in studies of growth, learning and behavior.
Much of this research has been based on well-characterized
noninbred stocks, but many recent studies are based on
inbred rats. It seems likely that genetically homogeneous
inbred rats will become increasingly important as research
tools. Old rats from different inbred strains develop
specific pathologic lesions, including important tumor
types. Two contrasting rat inbred strains have been
developed--one with low, the other with high, incidence of
dental caries. Special rat stocks have been developed for
study of hypertension--one that develops hypertension
spontaneously, and one whose hypertension develops only
with a salt-containing diet.

Other Species   Important genetic susceptibilities have
also been recognized in a variety of other mammalian
species. Mutant rabbits provide models of chondrodys-
trophy, adrenal hyperplasia, hemolytic anemia associated
with leukemia and scoliosis, among other diseases. Mutant
alleles carried by both pigs and dogs cause von Willebrandt's
disease. Other strains of dogs have congenital heart defects, similar to human anomalies; still others develop
rheumatoid arthritis or hemophilia. When armadillos are
infected with leprosy they develop lesions corresponding
more closely to human diseases than those to be found in
any other experimental mammal.

Here, as elsewhere, the full usefulness of scientific
reports describing these (and other) constitutional diseases in animals rests on the stocks being available in
the future. Recognition and long-term maintenance of
mutants in species not otherwise widely used in biomedical
research is an issue that merits careful attention.

Critical Issues

Inventory   Perhaps the most pressing need in the management of research collections is their coordination. By
their very nature, these collections are maintained by
persons with different interests and little communication.
Perhaps the first step toward responsible management is
to make an inventory. The major stock centers for well-
known model systems are readily identified, but most of
the smaller research collections are known to and used
by only a relatively small group of investigators. Moreover, many research workers--even though they are not
responsible for a stock center--possess unique genetic
materials; e.g., mutants collected for a particular purpose, specially constructed stocks, or wild strains
brought in for comparisons. One must be concerned, therefore, with maintaining current data both on collections
of uncommonly used species and on uncommon genetic variants
of commonly studied species. The newsletters of the major
systems serve these needs, but no comparable service is
available for the other materials.

Evaluation   Not only do we need a centralized inventory
of the collections, but the collections must be evaluated.
The curator of a collection is often not an appropriate
judge of the value of his stocks, even though--or because--
he is very much aware of their cost. The need for

evaluation arises from the cost of perpetuating collections
and the limited funds available for such things.

Continuity Because a large fraction of the research
collections are maintained by a few persons, or only one,
their continuity often depends on the health, fortunes,
or competence of a single investigator. If the principal
investigator becomes ill or goes unfunded or retires, the
materials assembled over a considerable career may be
discarded or fall under incompetent management and be
allowed to deteriorate.

Security Preservation of collections competes with other
aspects of research. The need for a fail-safe incubator,
a back-up electrical supply system, or a duplicate storage
facility must vie with the need for a new centrifuge or
microscope. A curator may undervalue his collection, at
least in its role as a public resource, or he may simply
be unaware of the best methods of managing it. A system
of inventory and evaluation should include advising curators about effective management practices.

Storage Technology Recent technological advances permit
much more economical and secure storage of many kinds of
biological materials. The precise application of these
techniques to particular species or strains, however,
often requires adjustment and trial, and the research
laboratories associated with minor collections are often
not equipped to adapt the methods to their materials.
More investigation is needed in the principles and practice
of germplasm storage.

Financial Support Financial support for collections of
genetic stocks is varied. Support of individual stock
centers ranges from solid-based grant or institutional
support to no support other than the investigators' personal resources. Furthermore, governmental support for
such centers has not kept pace with growth of research.
In some instances wherein governmental support has been
discontinued funds have been provided usually on a tem-
porary basis--by educational institutions or another
agency.

The lack of dependable support seems incongruous with
the importance of many of these collections as national
and international resources. The importance of the collections, and the need for funds for maintaining them,
should be brought to the attention of the public and
various funding agencies.

Yet we cannot maintain every stock center or resource.
It is difficult to determine just which resources should
be kept and which have least likelihood of being important
for research in the future. The National Science Foundation has asked the Committee on Maintenance of Genetic
Stocks of the Genetics Society of America to suggest guidelines for making these judgments.

Responsibility Genetic stock resources are for the most
part maintained by people who have research interest in
their collections and who have taken, for various reasons,
the additional responsibility of propagating and disseminating stocks for their own use and that of colleagues.
Under these circumstances, immediate availability of a
variety of stocks is an experimental advantage to the
stockkeeper. This and other advantages offset the burden
of finding funds to support maintenance of the stocks.

The greatest hazard attendant to individual supervision
of collections is lack of permanence. Upon the death or
sudden departure of the key individual, decisions as to
continuation may be made by scientists or administrators
who have no special insight into the stocks or the research involving them. Stocks are too often discarded
without due notification to scientists who need them.

The Committee on Maintenance of Genetic Stocks of the
Genetics Society of America recommends that an advisory
body be formed around each major stock center and that
these advisors come from outside the institution or agency
that harbors the stock. These advisors should be chosen
because of their expertise and interest in the particular
stocks, and would assume the responsibility of arranging
for the continuation of the stocks in the event of the
departure of the stockkeeper. These advisors could also
help the stockkeeper in determining what materials to add
or delete and in other matters pertaining to stock maintenance. Additional responsibilities could be delegated
to the advisors, their nature depending on the needs of
the stockkeeper and his institution.

ORGANISMS FOR RESEARCH IN FIELDS OTHER THAN GENETICS

Organisms collected from the wild are used in many types
of research, including cell biology, developmental biology,
molecular biology, neurophysiology, comparative endocrinology and physiology, ecology, ethology, and systematics
and evolution.

Research in molecular biology has been done mainly on genetically defined organisms, but some important contributions have come from molecular studies of plants and animals from natural populations. These include the crystallographic studies of whale myoglobin, studies on gene reiteration and gene amplification in amphibians, and studies of genome organization in sea urchin embryos.

The same general statement can be made about cell biology. Research in this area is done both on laboratory cultures and on organisms collected from the wild, but some of the more important work (e.g., that on chromosome structure and function) has been done on chromosomes of species collected from nature. Chromosomes of nematodes, insects, amphibians, and angiosperms have been particularly valuable as objects of study.

Research in developmental biology (embryology) traditionally has been carried out with organisms collected from nature. In fact, most of the important discoveries in this field have been made on material so obtained. Our knowledge of fertilization, the morphogenetic organization of the early embryo, embryonic induction, cell-cell interactions in tissue and organ formation, hormonal control of development, and the role of the nucleus in development has been obtained mainly from studies on organisms collected from the wild.

Fields of inquiry such as comparative endocrinology and physiology, ecology, ethology, systematics and evolution are almost totally dependent on natural populations of animals and plants.

Organisms in Ecosystems

Preservation of a wide variety of relatively undisturbed natural and managed ecosystems plays an important part in making it possible to detect and measure change, so that intrinsic factors contributing to natural change can be distinguished from humanly induced changes. An ability to do this will mitigate the common weakness in "before and after" studies of effects of environmental changes (e.g., damming a river or releasing heated power-plant effluent into the marine environment). At the present time, assessments of the "before" condition may be unrepresentative and observed changes may well include some that would have occurred even without imposed environmental disturbance.

Three kinds of intrinsic changes occur in ecosystems:

(1) Cyclical--at times very dramatic (e.g., the large fluctuations in population abundance in certain species of small mammals), but seem often to be related in some way to long-term persistence of the population; (2) Successional--may be slower than cyclical changes and thereby give an impression of stability; ultimately, however, they result in the disappearance of certain species (in managed ecosystems an effort is often made to hold the successional sequence at a subclimax stage); and (3) Stochastic--unpredictable by definition; the resilience of many ecosystems permits a partial or complete recovery to the previously prevailing stage. Above and beyond these intrinsic changes are the effects of external influences, chiefly those related to intensified human activity.

In order to assess the relative effects of these changes on natural and managed ecosystems, environmental surveillance techniques must be invoked. A variety of measures have been used. Surveillance may be undertaken at any organizational and trophic levels within an ecosystem, from the individual to communities. For example, the identities of species present can provide considerable information, if their biology and ecology are well understood. The realization that valuable environmental insights could be derived from a species list, provided tolerances of individual species and relationships between the species were better known, has led to the concept of indicator species. In practice, however, such species are usually highly opportunistic organisms that have short generation times, wide dispersal capabilities, and the ability to take advantage of disturbance by increasing their population size very rapidly. The sudden explosion of a pest species population is usually an indication of gross changes in the ecosystem. Early effects of pollution usually include the disappearance of rare species from the ecosystem, yet most environmental surveys and laboratory tests of pollutants cannot perceive or monitor these effects.

At the community level there are certain properties, such as diversity (the number of species and the distribution of individuals among species), that most nearly approach an ideal single measure in that they integrate the many individual and group contributions. Experienced biologists can make valuable assessments of data from single surveys by inspecting a list of species and their relative or absolute abundances. However, extensive surveys, in both space and time, produce such a mass of data that a summary, usually in the form of an index, must be

substituted. Costs of sample collection, sorting, and
identification are high compared with the cost of data
analyses. It is therefore important to make the fullest
possible use of the results, and different indices may
profitably be used for different purposes. Hellawell (1977)
has summarized the chief properties of a number of indices.

Marine Species

As noted earlier, the variability residing in populations
of marine species may be distributed in a continuous or
semicontinuous way over a large latitudinal gradient.
Effort to understand the meaning of the total genetic
variability for the survival of the species should therefore be intensified and a better understanding of the
degree of interconnection and interdependence between
semi-isolated populations sought. Marine and estuarine
environments include a number of ecosystems characterized
by extremely high levels of diversity, and many marine
species are themselves highly genetically variable. It
has already been argued that the only method of preserving
these high levels of genetic diversity is to preserve
natural ecosystems in sufficient number and variety, and
of sufficient size, to maximize the diversity of forms
that can be maintained, and that special attention should
be given to the most diverse systems: the tropical coral
reefs and the deep sea.

Considerable progress has been made in developing
laboratory techniques to maintain marine species of flora
and fauna (Kinne and Bulnheim, 1970; Smith and Chanley,
1975). Yet there are still many features of the life
history and biology of even relatively well-known commercial species that are not well understood, and this leads
to difficulties when the species are raised on a large
scale under hatchery conditions or are "farmed" under
natural conditions.

Research on the marine environment is conducted and
funded by a large number of agencies and organizations
having overlapping interests and priorities. In order to
carry out the task of preserving the greatest possible
diversity of germplasm, appropriate organizations must
assume responsibility for inventorying species, classify-
ing and studying ecosystems, designating areas that should
be preserved, promoting basic research on the biology and
ecology of marine species, promoting applied research on
marine species, studying acute and chronic effects on

organisms and ecosystems of a wide variety of pollutants, and long-term monitoring of trends and changes in the marine environment. At the present time each of these responsibilities is assumed by a number of the following: universities; private research organizations; commercial companies conducting environmental studies and aquaculture projects; National Science Foundation; National Oceanic and Atmospheric Administration (including National Marine Fisheries Service and Office of Sea Grants); National Institutes of Health; Department of Energy; Environmental Protection Agency; Office of Naval Research; Bureau of Land Management; U.S. Department of the Interior (for example, the Bureau of Indian Affairs funds certain aquaculture projects); U.S. Geological Survey; Council on Environmental Quality; and municipal, county, state, and regional authorities.

There is much reason to feel that this multiplicity of agencies should be condensed into a single authority; on the other hand it has been argued that there are certain valuable safeguards in having a number of agencies funding projects related to their special missions.

Procedures for funding research generates certain difficulties. For example, when the total amount of money available for research becomes limited, there is a tendency to shift the emphasis from basic to applied research, to short-term rather than long-term projects, and to work that is thought to promise quick payoff. Most of the agencies fund projects on a year-by-year basis; few research grants commit funds for periods greater than 2 years. Yet certain kinds of work require a more sustained effort and some assurance of funding for a longer period (e.g., long-term ecological measurements and genetic studies).

Again, overlapping authority permits important, but perhaps unfashionable, research to go unsupported because it is not the special responsibility or charge of a particular agency. One such area is taxonomy, where for many groups of organisms there is but one expert in the world, or perhaps none. Yet unless the species in question is correctly identified, access to the literature on the genetics, physiology, behavior, ecology, or evolution of that species is hindered and it is therefore not possible to gauge the reliability of the information. For example, many baseline environmental surveys are based on a taxonomically very weak inventory and classification. Thus, the future value of such studies as reference points is highly doubtful. Yet the amount of educational and

research support for training students in taxonomy reflects a general lack of research support for professional
taxonomists.

Other areas of research suffer not so much from general
neglect as from rapid changes in agency responsibility.
Studies in marine genetics constitute one such area. All
agencies concerned with marine research seem to be skepti-
cal about the intrinsic value of genetic studies. Primary
responsibility for genetic studies has not been clearly
expressed or allocated, rather it has been transferred
from one agency to another several times.

Macroscopic Organisms in Culture

Organisms now carried in culture were in the past often
harvested on a continuing basis from wild populations.
The trend toward maintenance under controlled conditions
has occurred for two reasons: first, man's impact on the
habitats of these organisms is thought to have reduced
the size of the available populations; and second, certain
research benefits from a greater uniformity in the genetic
background of the materials under study.

Certain classic organisms have long been used in fundamental studies in cell and developmental biology. One
thinks particularly of amphibian eggs and embryos, and
sea urchin eggs.

Amphibians   In the United States many leopard frogs (*Rana
pipiens*) are used in teaching laboratories, and many bull-
frogs (*R. catesbeiana*) as food. Up to now, virtually all
of these have been collected from nature. So long as
their habitats remain intact, populations of these species
withstand both extensive collecting and occasional natural
reductions. For example, where a long Indian summer is
followed by freezing rains, leopard frogs in Vermont may
not migrate to Lake Champlain to hibernate and may incur
large losses over the winter, but the population recovers
completely in about 3 years (Nace, 1976). Also, some
experiments have been performed in which the frog populations of given lakes were almost completely removed; again
the populations recovered in about 3 years (Nace, 1976).
Thus amphibian populations show a remarkable capacity to
recover from extensive losses so long as their habitats
are preserved.

This situation appears to have changed during the past
several years. Losses and changes in habitat, and perhaps

losses from excessive collecting, have brought about reduction in the supply of the commonly used species.
Several measures may be taken to combat this trend: habitat preservation; enactment of laws governing the collection of amphibians; preservation of migration routes
between foraging and hibernating areas (e.g., by construct-
ing highway underpasses); and commercial culture of
commonly used species.

No one has yet succeeded in large-scale commercial
breeding of amphibians, although efforts are being made.
G. W. Nace (personal communication) states that he has
about 15,000 frogs in his amphibian facility. He and his
associates have developed new population cages, feeding
devices, nonliving foods, and better methods of carrying
the animals through metamorphosis, all of which have led
to improved efficiency and productivity. However, Nace
estimates that he is still several years away from ac-
complishing a commercially feasible operation. D. D.
Culley (Lousiana State University) seems to be in about
the same situation.

Of all the anurans, *Xenopus* appears the easiest to
culture. It is still available from natural populations
in large numbers, but import restrictions may affect
access in this country.

One of the simpler aspects of maintaining amphibian
germplasm concerns genetic stocks. These stocks, important in several types of biological research, are maintained in a few stock centers and in research laboratories.
The long-range objective is to ensure that mutant stocks
are preserved. If stocks are lost, it might be very difficult or impossible to produce them again.

Amphibian colonies are usually maintained by natural
breeding; occasionally artificial insemination is used.
Methods for routine preservation of sperm, eggs, and early
embryos by freezing are not available. The most commonly
used species of amphibians produce large numbers of eggs
at each spawning, and in captivity may be induced to spawn
every 2 or 3 months. However, generation time is long,
ranging from about 7 months to more than a year. For
genetic studies it would obviously be desirable to search
for amphibian species with generation times considerably
shorter than this.

Sea Urchins   Sea urchin eggs have also been used in fundamental studies in cell biology. Yet it was only a few
years ago that the populations of *Arbacia punctulata* off
the coast of New England declined dramatically. It was

then apparent, first, that biologists did not fully understand the reasons for the decline and, second, that it
would be difficult to substitute material from other species for comparable studies. Since then, other species
of sea urchins have been exploited more heavily as a research resource, but even some of these populations (e.g.,
Strongylocentrotus purpuratus) may be threatened by commercial exploitation. This has led R. Hinegardner and
others to search for reliable methods for culturing species
of sea urchins under laboratory conditions. These techniques and others applicable to a variety of marine organisms will be summarized in a forthcoming report by the
Committee on Marine Invertebrates, ILAR, National Research
Council.

Other Species   Other species of harvested organisms--the
squid, several species of aplysiids, the lobster, and the
horseshoe crab--have long provided the material for fundamental studies in neurobiology. Some of these continue
to be obtained solely from wild populations, but the
Division of Research Resources, National Institutes of
Health, has funded studies on the culture and life-history
attributes of squid and a number of closely related species
of Pacific aplysiids.

Various other organisms are in specialized collections
important to basic research. A few examples will suffice:
Caenorhabditis elegans (behavior studies); butterflies,
Tribolium (population biology); and *Chironomus* (gene
regulation). Others are raised for toxicity testing of
pollutants.

Guidelines for toxicity testing of pollutants (Stephan,
1975) recommend the use of large or relatively robust
species, the primary criterion for this purpose being that
they should be readily available or commercially important.
Most such species are in fact obtained from commercial
suppliers, and only in some species of fish are hatchery
stocks, of known genetic backgrounds (for example, the
fathead minnow), used in such tests. Some pollutant
studies on marine invertebrates have used species that
have been inbred over several generations, but in most
cases little is known about the effects of inbreeding *per
se*.

Studies of marine pollution have increasingly shown
that the pelagic larval stages are likely to be highly
sensitive and therefore of special value in studies of
the effects of pollutants. However, the parents of the
larvae are, in almost all cases, taken directly from wild

populations; hence genetic differences in susceptibility to pollutants are not taken into account.

One broad-scale study, recently initiated by the Environmental Protection Agency, will use the mussel *Mytilus edulis* as a monitor of pollutants. This species is widely distributed along the coasts of Europe and the United States and the post-larval stages are sessile and relatively long-lived. Thus populations provide evidence of trends in the environment, and individuals act as indicators of environmental effects.

Microorganisms in Culture

Most of the cultures of microorganisms in the United States are in specialized collections assembled with a view to public-health needs, agricultural or industrial needs, or for use in research. It is often difficult to distinguish microbial collections of practical importance from basic research collections, as illustrated by the collections of algae and bacteria used in studies of photosynthesis. These may be assembled in governmental or industrial laboratories addressing specific practical problems or in university research laboratories studying mechanisms of photosynthesis at the molecular level. Much research of basic significance occurs in practical laboratories, and much work of practical value comes out of the basic research laboratories. The same can be said of collections of nitrogen-fixing organisms, where the results of basic research are applied promptly to the solution of practical problems of immediate concern to world agriculture. The collections of organisms used in basic medical research stand in a similar relationship to those used in medicine and public health, as do many of those collections used in basic biochemical, pharmacological, and physiological research.

A few research collections consist of specially constructed stocks or wild collections useful mainly to an individual investigator; some consist of large, well-cataloged resources valuable to a considerable community of scholars. The extremes pose no problems. The "personal" collections cannot ordinarily be maintained once the "person" disappears from the scene. The obviously useful public collections will surely be preserved by some means.

The crux of the problem lies with collections that fall between these two extremes, given limited funds for preservation. A central decision-making organization composed of qualified scientists representing broad areas of expertise is required so that priorities can be set and hard choices made. The responsibilities of such a group should include not only the selection and perpetuation of good collections but also the recognition and support of new collections of potentially significant organisms not presently backed by large numbers of users.

One laudable trend, which should be encouraged and supported, is the establishment of broadly specialized central collections such as have been set up in recent years by persons responding to national needs. The Algae Collection at the University of Texas, Austin, assembled by R. C. Starr and now supported by the National Science Foundation, is an important resource operating at an effective level.* The International Collection of Phytopathogenic Bacteria established at the University of California, Davis, by M. P. Starr is the result of another effort to make widely available a comprehensive collection in a broadly specialized field. Although an extensive and valuable collection has been assembled, this center is unable to respond to more than a small fraction of the requests received, because of insufficient operating funds (M. P. Starr, personal communication). It is further threatened by the impending retirement of its curator.

The collection of the Anaerobe Laboratory, located at the Virginia Polytechnic Institute and State University, is another example of a valuable, broadly specialized collection. Its some 22,000 strains are heavily utilized, both nationally and internationally, by laboratories involved in human medicine, veterinary medicine, food production, pharmaceutical production, and the production of energy by microorganisms.

A system that incorporates such collections, each in the hands of competent curators, could do much to reduce the inefficiency and redundancy of the present situation. After areas of coverage had been firmly established and widely recognized, and curatorial expertise and continuity had been assured, many small specialized collections

---

*The Algae Collection is arranging to acquire the more valuable stocks from the collection of L. Provasoli, of the Haskins Institute and Yale University, who has retired. These include unique pure cultures of marine phytoflagellates and marine invertebrates (Provasoli, 1976).

could be eliminated or consolidated, and valuable stocks
from small collections could be saved, when endangered,
by placing them in the central collections.

The system of broadly specialized national culture
collections for microorganisms of economic importance has
been adopted in the United Kingdom and the Commonwealth
nations. An effort to determine how well this system has
worked out in those countries would be warranted.

Ciliated Protozoa

The ciliated protozoa suggest an informative case history
in germplasm management. They do not represent a topic
of sufficient magnitude for extensive consideration here,
but they may represent a class of problems. About 6,000
species have been described, but the more intensively
studied species are found to have been previously "under-
classified" (Sonneborn, 1975; Nanney and McCoy, 1976).
Little systematic attention has been given to many major
groups. A conservative extrapolation suggests that the
number of isolated gene pools is at least 10 times greater
than the number of named species; it may be as much as 100
times greater.

Cell Cultures

Cell strains from a wide variety of animals are used in
biomedical research. For example, until very recently
many nonhuman primates were killed in order to obtain
kidney cell cultures for propagation of viruses in vaccine
production. Most mammalian cell lines are readily preserved (with adequate protective agents) in liquid nitrogen. Cells of animals that die in zoos and at importation
points are sought by geneticists, those doing research on
aging, and other scientists. Only one center seeks to
collect cells for its own research program. About 250
strains of various species are on hand in a collection
maintained by the San Diego Zoological Society. In a few
other laboratories strains of a few species are collected,
but with no commitment to continuance.

Almost all of these cells are mammalian. Ten cell lines
of eight species of fish have been preserved for diagnostic
purposes by the U.S. Fish and Wildlife Service's Eastern
Fish Diseases Laboratory in Virginia, indicating that
preservation of fish cells is feasible. No lines of bird
cells are known to us.

A large repository of "human genetic mutant cells" is maintained by the Institute for Medical Research, Camden, New Jersey, sponsored by the National Institute of General Medical Sciences (NIGMS), NIH. It is composed of well-characterized genetic mutants of human disease, chromosomal errors, tumors, and the like and exists as frozen vials of cell strains or cell lines. The bank, operated by scientists, periodically publishes a complete list of its holdings (third edition, August 1976). It also publishes descriptions of strains in scientific journals. The strains are available for purchase by qualified scientists.

In addition, the American Association of Tissue Banks, Rockville, Maryland, had its inaugural meeting in May 1977. The purposes of the association are stated in its Constitution and Bylaws as follows:

1. To promote scientific and technical knowledge concerning the procurement, processing storage, transplantation and evaluation of cells, tissues and organs, for clinical and research uses; hereinafter the term "tissue" shall include cell products, cells, tissues and organs;
2. To encourage the voluntary donation of cells, tissues and organs for clinical or research purposes;
3. To make available through regional tissue bank programs a safe, adequate and economical supply of tissues and organs for both clinical and research purposes;
4. To inspect and certify repositories for cells, tissues and organs used for clinical or research purposes; and
5. To establish Codes and Standards for cells, tissue and organ preservation used for clinical or research purposes.

It may well be that this organization will ultimately coordinate the efforts of numerous scientists in the collection and safeguarding of cells and tissues.

The American Type Culture Collection is a grant- and fee-supported private organization that collects, stores, characterizes, and ships a wide variety of animal and human cells, viruses, microorganisms, and so on. Its first catalog of cell lines was issued in 1975. Few wild animal cell lines are available through this collection, whose input depends on donors. The cost of cells to investigators is considerable, and it is difficult to understand how it is decided which cells to collect and maintain.

7 CRYOBIOLOGICAL PRESERVATION

An alternative to preserving germplasm by maintaining reproducing colonies in the wild or in the laboratory is to preserve it in a dormant state. The effectiveness of such an approach depends on the variety of cells in which dormancy can be reversibly induced, and it depends on the economics and reliability of the preservation techniques and on the institutional arrangements that are used.

The traditional procedures for storing or banking germplasm in the dormant state have been low temperatures, low water content, or a combination of the two. They have been used widely to bank microorganisms, cattle sperm, plant seeds, and animal tissue-culture cell lines. Other methods have been described from time to time, such as preservation of fungi under mineral oil, but they are mostly applicable to only a relatively few organisms. Here we are concerned with a rather special aspect of preservation in the dormant state--induced long-term dormancy achieved by freezing, drying, or both. This state has been termed "cryptobiosis" by Keilen (1959) and defined by him as "the state of an organism when it shows no visible sign of life and when its metabolic activity becomes hardly measurable, or comes reversibly to a standstill."

Closely coupled with the process of preserving germplasm in a dormant state is the need to resuscitate it when needed. Resuscitation, of course, includes consideration of the scientific and technical aspects of thawing frozen cells and with the rehydration of dehydrated or freeze-dried cells. The term can be extended to include the eventual utilization of the preserved materials, e.g., preserved viviparous embryos require the existence of suitable foster mothers or of techniques that permit complete embryological development *in vitro*. Again, to use

79

preserved plant-tissue cultures there must be techniques
for inducing the regeneration of the full plant from somatic cells in culture.

CRYOBIOLOGICAL PRINCIPLES

Freezing

The preservation of cells by freezing subjects them to a
number of sequential steps any one of which is potentially
lethal. These steps are: (1) the collection of the
material and, for animal cells and some plant materials,
its transfer to media that generally must contain molar
concentrations of nonphysiological protective solutes
(e.g., glycerol or dimethyl sulfoxide); (2) freezing *per
se* to temperatures below -130°C; (3) low-temperature
storage; (4) warming and thawing; and (5) removal of the
protective solute and return to normal physiological conditions.

Steps (2), (4), and (5) are most likely to induce
injury. The cooling rate in step (2) is especially critical. Cells must be cooled at controlled rates, the numer-
ical values of which can differ by as much as 1,000-fold,
depending chiefly on the size of the cells and their permeability to water and on the protective additive used.
The mechanistic effects of cooling rate are becoming better understood, i.e., too high a cooling rate results in
death from the formation of ice crystals within cells and
their growth during warming, whereas too low a cooling
rate causes death from the chemical consequences of the
concentration of solutes during freezing or from osmotic
forces operating during cooling and subsequent warming
(Mazur, 1970, 1977; Farrant *et al.*, 1977b; Meryman *et al.*,
1977).

The effects of step (4) depend in a rather complex way
on the prior cooling rate. Injury during warming results
from growth of intracellular ice if cooling has been fast
enough to induce it; rapid thawing minimizes this effect.
Injury during warming also results from osmotic forces
originating from events that occurred during slow cooling,
forces that seem to be minimized by slow warming. Osmotic
forces similarly play a role in step (5), the removal of
the protective solute.

Step (3), low-temperature storage, causes no difficulty
if the storage temperature is sufficiently low. "Sufficiently low," in this context means at least below -130°C,
the glass transition temperature of water, or preferably

-196°C, the temperature of liquid nitrogen. At -196°C, there is good reason to expect that most of a population of stored cells will remain viable and unchanged for centuries or, more likely, for a millennium. At -196°C (and probably at -130°C as well), no known, biologically relevant, thermally driven chemical reactions can occur. The only sources of damage are photophysical events: ionizations produced by background radiations and high-energy cosmic ray protons. It has been calculated that 200 years would have to elapse to accumulate sufficient radiation to kill 63 percent of early mouse oocytes, one of the more sensitive mammalian cells, and some 3,000-20,000 years would have to elapse to kill 60 percent-90 percent of a population of "average" mammalian cells (Mazur, 1976; Ashwood-Smith and Friedman, 1977). So far as the induction of mutations is concerned, because less than 1 percent of the spontaneous mutation rate at physiological temperatures is due to background radiation, at least in mouse germ cells (Russell, 1963), and because background radiation should be the only factor contributing to mutational alterations at -196°C, the mutation rate at that temperature should be considerably lower than the spontaneous rate, even though biological repair is precluded at -196°C (Mazur, 1976; Ashwood-Smith and Friedman, 1977).

These theoretical considerations are consistent with experimental evidence. Although biological decay at -196°C is occasionally claimed, the consensus is that these observations are the result of artifacts and that no decay occurs at -196°C. Experiments now in progress subject mouse embryos to 100 times background γ radiation at -196°C. As expected, no adverse effects have been detected after 2 years of irradiation (Lyon et al., 1977). On the other hand, at temperatures above -80°C, there are many well-documented cases of rapid decline (days to months) in cell viability (Meryman, 1966a). Here, too, there are sound physical reasons for expecting such declines.

An important question regarding the use of low temperatures to preserve germplasm is the extent to which freezing either induces mutations or acts to select preexisting variants in the population. Although the question has not been studied adequately, available information indicates that mutagenesis is nil. There is not only direct experimental information on microorganisms (Ashwood-Smith and Friedman, 1977), but several experiments on the freezing of DNA itself have shown no discernible effects (Shikama, 1965; Ashwood-Smith et al., 1972; Ashwood-Smith and Friedman, 1977).

Nor does freezing appear generally to select preexisting variants in a population, a conclusion based both on direct examination in bacteria (Ashwood-Smith and Friedman, 1977), and on crude "epidemiological" evidence from the experience of repositories of frozen cell lines and from the large-scale use of frozen bull spermatozoa in the cattle industry. There are, however, a few documented instances of selection. Wild-type T4 bacteriophage and an osmotic resistant mutant differ greatly in their response to cooling rate; and in a mixture of the two types, cooling rates can be adjusted so as preferentially to select the osmotic resistant form (Leibo and Mazur, 1970). Similarly, germinated spores of the fungus *Neurospora* are much more sensitive to freezing than are nongerminated spores. Accordingly, if a population of spores is suspended in minimal growth media and frozen, the freezing will select in favor of auxotrophic mutants that are incapable of germinating in the minimal medium (Leef and Gaertner, 1975). Some cell populations (e.g., mammalian marrow cells and white blood cells) are physiologically and developmentally heterogeneous and differ in their cooling rate optima sufficiently to permit the enrichment of a given type by freezing (Austin, 1973; Farrant *et al.*, 1977a). Selection of preexisting variants can be minimized, of course, if freezing techniques are such as to produce overall high survivals.

Finally, there is the question whether the processes in freezing cause nonheritable and nonlethal morphological and physiological changes, and whether the freezing of germ cells and early embryological stages has teratogenic effects. Although the issue has not been studied in depth, it is clear that: (1) in most cases cells or organisms that survive freezing are indistinguishable morphologically, physiologically, and biochemically from the normal population, and there are no reports of teratogenic effects from freezing of sperm or early embryos (Whittingham *et al.*, 1972; Maurer *et al.*, 1977); (2) on the other hand, there are some well-documented cases where freezing does induce nonlethal alterations. Frozen-thawed sperm, for example, at times retains motility while losing its ability to fertilize.

The chief concern in using freezing to preserve germplasm, therefore, is not the sheer maximum attainable length of low-temperature storage, nor the induction of genetic or physiological defects in the survivors, but is to understand the events in freezing well enough to allow desired cells to survive both cooling to -196°C and the later return to normal physiological temperatures and media.

Freeze-Drying and Dehydration from the Liquid State

Freeze-drying (lyophilizing) consists of first freezing
cells and then removing frozen water from them at low
temperatures by sublimation *in vacuo*. A major fraction
of the cellular dehydration actually occurs during the
initial freezing, as the liquid cell water is converted
to ice. The chief difference between freezing and freeze-
drying is the ultimate degree of dehydration. Freezing
effectively removes the approximately 90 percent of the
cell's total water content that is "unbound" but it does
not remove the residual bound water. Freeze-drying removes most of the latter as well (Meryman, 1966b; Mazur,
1968). So also do processes involving dehydration from
the liquid state at above zero temperatures.

This difference between freezing on the one hand and
freezing-drying or dehydration from the liquid state on
the other hand probably explains why the latter two processes are considerably more deleterious to living systems.
Generally the only cells capable of withstanding freeze-
drying are bacteria and the spores and other dormant forms
of fungi that regularly withstand dehydration in nature.
The ability to survive freeze-drying is often critically
dependent on the final water content. Drying to too low
a water content can be immediately lethal; drying to too
high a final water content causes rapid killing during
storage of the dried product.

An early rationale for freeze-drying was that long-
term preservation of cells would obviate the need for
refrigeration. But this is not so; storage of freeze-
dried cells at 4°C to -18°C is required to prevent rather
rapid decline in viability (Fry, 1966).

There are well-documented instances in bacteria and
plant seeds in which freeze-drying and air-drying induce
mutations and chromosomal aberrations (Roberts, 1975;
Ashwood-Smith and Grant, 1976). Roberts (1975) has calculated that the mutations accumulating in dried barley seeds
stored under conditions that have led to a 50 percent loss
of viability are equivalent to those induced in fresh seeds
by 10,000 rad of X-rays. There is also a potential danger
of selection, since in many cases only small percentages
of cells in a population survive dehydration.

Although mutagenesis and selection are potentially
troublesome, a number of culture repositories routinely
freeze-dry and store microorganisms without apparent
problems of heritable differences.

It is entirely possible that procedures could be

developed for successfully freeze-drying a broader spectrum
of cells, but the outlook is much less encouraging than
for freezing, per se. Pragmatically the only advantage
of freeze-drying over freezing is that the former does not
require exceedingly low temperatures for long-term storage.
But the cost of long-term cryogenic storage of frozen cells
is scarcely prohibitive, and it is certainly technologically
feasible to design adequate sensors, alarms, and backup
systems to detect and rectify failures in the primary refrigeration systems.

STATUS OF CRYOBIOLOGICAL PRESERVATION OF GERMPLASM

Germplasm can be preserved in four ways: (1) by preserving the whole plant or animal; (2) by preserving somatic
cells derived from the plant or animal; (3) by preserving
individual germ cells or their direct progenitors (e.g.,
sperm, ova, pollen, spermatocytes, or oocytes); and (4)
by preserving embryos. Methods (2), (3), and (4) have in
common the requirement that procedures must exist for re-
converting the preserved cells into reproductively competent plans and animals.

Prokaryotes

Most bacteria can be frozen such that they have high
enough survival rates to permit the establishment of sub-
cultures from the thawed progeny (Mazur, 1966). Some
blue-green algae survive freezing well; others do not
(Holm-Hansen, 1963). The basis for the difference is not
known.

Plants

Whole Plants Most reproductively competent mature plants
do not survive freezing to -130°C or below. This is true
of fungi, mosses, ferns, and tracheophytes (Levitt, 1956,
1964; Mazur, 1968, 1969; Weiser, 1970), but there are
exceptions. Some higher plants undergo a hardening process
in nature that permits them to survive temperatures as low
as -60°C (Levitt, 1956, 1964; Weiser, 1970). Some vegetative fungi, such as yeast, and many algae can survive
freezing to -196°C under appropriate conditions (Mazur,
1966, 1968).

Somatic Vegetative Cells   The above statements on whole
higher plants apply for the most part to the preservation
of differentiated plant tissues, such as leaves, stems,
and roots, although here there are numerous cases of
hardened woody tissues surviving to -196°C (Burke *et al.*,
1976). In contrast, there have been recent reports of
the successful freezing of cultures of plant-tissue cells
in which procedures similar to those developed for animal-
tissue cultures were used. To date, cultures from about
10 species of plants have been frozen with varying success (Dougall and Wetherell, 1974; Nag and Street, 1975;
Bajaj, 1976; Towill and Mazur, 1976; Seibert and Wetherbee,
1977). The potential for the preservation of germplasm
lies in developing procedures for the generation of whole
mature plants from apical meristems or other relatively
undifferentiated somatic cells. Full regeneration from
tissue culture can now be achieved in carrot and tobacco
(Steeves and Sussex, 1972), and undoubtedly procedures
for the freezing and regeneration of other species will
follow in due course. But achieving broader success may
not be easy; certain forms, for example, are exceedingly
sensitive to freezing (Towill and Mazur, 1976).

Reproductive Bodies   Spores, cysts, pollen, and seeds
that withstand air-drying in nature usually withstand
freezing, if they are air-dry at the time of freezing.
Full hydration, or germination, prior to freezing often
causes them to lose their resistance (Ching and Slabaugh,
1966; Mazur, 1968b). Some spores, pollen, and seeds have
not been successfully frozen. This latter result may be
correlated with their possessing high water content in
nature and with their relatively low resistances to desiccation.

   According to Roberts and Ellis (1977), the International
Board for Plant Genetic Resources has recommended that most
seeds be stored at -18°C. Roberts and Ellis present data
for barley showing that 1 day's storage at +40°C equals
about 200 days at -20°C. Although -20°C may also be a
satisfactory storage temperature for seeds other than barley, it will be costly and time-consuming to establish this
fact. From the fundamental cryobiological consideration
for hydrated cells, -20°C is an undesirable temperature,
at least far less desirable than $\leq$-100°C (Mazur, 1966).
True, storage at $\leq$-100°C is far more expensive than storage at -20°C, but to the cost of the latter must be added
the considerable costs of periodically regrowing stored
seed (Weiser, personal communication).

Animals

Intact Organisms, Except Embryos   Instances of intact
animals being frozen are restricted mostly to the protozoans, although there are sporadic reports of successful
freezing of metazoans (e.g., rotifers, nematodes, and
insects) (Asahina, 1966; Koehler and Johnson, 1969; Haight
et al., 1975). No vertebrate has ever survived freezing
to more than a few degrees below 0°C. Even among the
protozoans there are numerous forms that either do not
survive freezing (e.g., *Amoeba proteus*) (Mazur, 1966) or
in which only small percentages survive even under optimal
conditions (e.g., *Paramecium*) (Simon, 1971). The outlook
for successful freezing of the larger, more complex metazoans is bleak, partly because they are composed of cells
of widely differing characteristics and partly because
increasing size itself introduces formidable barriers.

Somatic Cells   Several types of individual cells (e.g.,
erythrocytes, lymphocytes, bone marrow, and myocardial cells)
and a few organized tissues (cornea and skin) can be frozen and retain high viability. But most mammalian tissues
and nearly all mature mammalian organs do not survive
freezing to below -20°C (Pegg, 1970; Karow *et al.*, 1974).
On the other hand, fetal organs (heart and pancreas) have
recently been frozen successfully to below -75°C (Mazur
et al., 1976; Rajotte *et al.*, 1976), and there are encouraging developments in efforts to freeze mature organs.
    Among the more successful examples of freezing in
animal cells have been tissue-culture cells. Most lines
can be frozen so as to yield high enough survivals after
thawing to ensure successful subculture (Pegg, 1970;
Coriell, 1976). At least two major repositories of frozen
cell lines exist in this country: the American Type Culture Collection and the Human Genetic Mutant Cell Repository.
    Unfortunately, the ability to preserve animal somatic
cells does not provide a method for the preservation of
animal germplasm in a form capable of reproducing itself.
Except for a few instances in lower forms, it is not now
possible to generate a complete animal from a collection
of its somatic cells and, unlike plants, it is not at all
clear when, if ever, it will be possible to do so.

Animal Germplasm   In the foreseeable future, then, the
preservation of animal germplasm by low-temperature storage will entail the preservation of gametes and early
zygotes.

• The preservation of bull spermatoza in 1950 (Smith and Polge, 1950) is said by some to have initiated modern cryobiology, and frozen sperm from highly selected bulls is now used in the artificial insemination of a large percentage of dairy and beef cattle. The sperm of a wide variety of other mammals has also been preserved by freezing. Those species in which frozen sperm has 50 percent or more of the fertilizing capacity of normal sperm include: bull, boar, stallion, ram, goat, and probably man. Those in which more than 30 percent of the sperm remain motile after freezing, but in which the fertilizing capacity has not been tested include: camel, moose, deer, bison, llama, yak, monkey, bear, chinchilla, dog, bighorn, and other wild sheep (Graham, 1973). Chicken and turkey sperm is viable after freezing, but both motility and fertilizing capacity are rather low (20-30 percent) (Graham, 1976). For several species of trout and salmon, fertilizing capacity of frozen sperm has proven to be high; for cod, moderate. Carp sperm is motile after freezing but has lost its capacity to fertilize (Horton and Ott, 1976). We have little information on sperm from invertebrates.

Although there seem to be no fundamental obstacles to the freezing of sperm from species not yet studied, the task will not necessarily be simple or routine. Over 20 years elapsed between the successful freezing of human and bull sperm and the successful freezing of ram and boar sperm. Empirical techniques developed for the former did not work for the latter, and many modifications failed. Neither the cause of freezing injury nor the basis of protection by various solutes is well understood.

• The successful freezing of sperm in 1946-1949 led to attempts with mammalian ova. Success, first achieved for mouse embryos in 1972 (Whittingham et al., 1972), derived from fundamental cryobiological and reproductive physiological information that had emerged in the preceding few years. Since 1972 embryos of rabbit, cattle, sheep, goat, and rat have been frozen and have yielded normal viable offspring when thawed and transferred to foster mothers (Muhlbock, 1976; Ciba, 1977). The Jackson Laboratory in the United States, the MRC Laboratory Animal Centre in the United Kingdom, and others are now beginning to use the technique as an approach to the long-term preservation of potentially valuable mutant lines of mice, especially those not used on a day-to-day or month-to-month basis.

Whether early embryos from mammals, other than those cited above, will now be frozen easily is uncertain. Early

stages of cattle embryos (1 to approximately 64 cells, cannot yet be frozen successfully, and pig embryos will not
even survive chilling to 0°C (Ciba, 1977).

We know of no instance where an ovum or embryo from
animals with large yolk-bearing eggs has survived freezing.
At least two reasons suggest themselves: (1) only in the
last few years has the successful freezing of cell aggregates been achieved (the obstacle has been that the permeability of cells to water and to solutes has a major
influence on their fate during freezing, and permeability
in turn is influenced greatly by cell surface-to-volume
ratio); and (2) oviparous eggs, having been exposed to
nonhomeostatic conditions, seem to have evolved protective
mechanisms, one of which is an outer layer that has exceedingly low permeability to water and solutes.

ISSUES MERITING FURTHER STUDY

In some cases the techniques empirically developed for
sperm have worked satisfactorily for other cells (e.g.,
many animal-tissue culture cells) and are in use today.
In other cases the original techniques did not work at all
well for other cells, not even for sperm of other mammalian species, and the derivation of successful procedures
has involved tedious empirical testing at various levels
of the several variables involved. In still other cases
both the original empirical techniques and subsequent
empirical modifications have failed. In a few of these
last cases (e.g., mammalian embryos) success finally has
been achieved because partial understanding of the fundamental aspects of low-temperature biology emerged in parallel with empirical studies. In other cases the level of
understanding is still too primitive to suffice. Success
in realizing the potential of cryobiological preservation
will depend in large measure on the skill with which the
techniques are applied to a wide variety of species and
cell types by investigators differing greatly in their
interests and in their knowledge of cryobiology.

The following seem among the more important questions
that need to be answered in order to expand the potential
of low-temperature biological techniques.

• What are the physical-chemical causes of slow freezing injury (e.g., concentration of electrolytes, osmotic
forces, and thermal effects)?
• Where do the critical lesions occur and what is their
molecular nature? There is increasing evidence that

membranes are more sensitive to injury than other organelles but, except for chloroplast thylokoid membranes (Steponkus *et al.*, 1977), the nature of the lesions remains largely unknown.

- What is the molecular basis for the action by which certain solutes protect against freezing injury?
- What are the permeability characteristics--and temperature dependences of permeability--of the cells in question, both to water and to solutes? Knowledge of the permeability characteristics has already proved a powerful asset in quantitatively predicting the optimum approaches to the freezing of several biological systems (Mazur, 1970, 1977; Leibo, 1977).
- Can the "scale-up" problem be resolved or ameliorated? As noted above, this issue relates to the fact that water and solute permeation depends on the surface-to-volume ratios of cells or cell aggregates, and these ratios decrease and become increasingly troublesome with increasing cell size: The larger the cell, the more slowly it must be cooled to avoid intracellular ice; the more slowly it has to be cooled, the more critical becomes the need for high concentrations of appropriate protective solutes; but the larger the cell or cell aggregate, the longer it takes for the solute to permeate into the interior of the aggregate; and, finally, the higher the concentrations of additive within cells or within cell aggregates, the greater becomes the likelihood of chemical toxicity from the solute and the more formidable become the osmotic problems associated with its removal after thawing.

Apart from questions such as the above, which apply to cells in general, there are sets of questions related to particular classes of cells. Answers to these more specific questions may well contribute not only to achievement of low-temperature preservation of the particular cell types but also to an understanding of their fundamental biology. For example:

- Why do sperm or ova from different mammalian species differ so strikingly in their susceptibility to freezing or even to chilling injury? Why is this sometimes also the case for different stages of embryos or for closely related cells from the same species?
- Can ova and embryos from oviparous and ovoviviparous animals be successfully frozen? If so, they will probably have to be made reversibly much more permeable to solutes and water. Can this be achieved?

90

• Some higher plant cells seem distinctly more sensitive to freezing than animal cells. Does this sensitivity
arise from the existence of a rigid cell wall? Is it
associated with interactions between the cell wall and the
protoplast? Or is it due to the presence of a large vacu-
ole in many plant cells?

RESUSCITATION

Germplasm that is preserved by dehydration or low-temperature
storage must, when needed, be converted into a form that
can reproduce. This conversion, or resuscitation, raises
a number of intriguing and significant questions.

Animal Germplasm

Ova and Embryos   It should not be difficult to resuscitate
frozen embryos of oviparous and ovoviviparous animals, provided ways can be found to freeze them successfully. But
the resuscitation of ova and embryos from viviparous animals (some of which can be frozen now) is far more complex.
The most direct approach is to transfer embryos to the
reproductive tract of foster mothers. Successful transfer
requires that the foster mother be in a suitable stage of
pseudopregnancy; this in turn requires knowledge of the
estrus or menstrual cycle of the foster mother. Although
such information is available for most laboratory and
domesticated species, much less is available for wild animals, even those in zoos.
    Cases arise where a nation needs to obtain nonindigenous
germplasm but is unable to do so because of restrictions
on the exportation and importation of animals, a matter
discussed elsewhere in this report. One route to importation of germplasm is to import frozen embryos. However,
the importation of embryos will be of no value if, because
of quarantine restrictions, there are no females of the
species to serve as foster mothers. Similarly, low-
temperature preservation of embryos has been suggested as
a method for preserving the germplasm of certain endangered
species. But again the preservation will not be helpful
if there are no females of those species left to act as
foster mothers.
    Perhaps the least speculative of the candidate tech-
niques is the rearing of embryos of the desired species
in the uterus of a closely related genus or species. This

approach necessitates an understanding of the nature of the interspecies barrier to *in vivo* embryological development. More speculative, but still in the realm of possibility, is *in vitro* embryological and fetal development. In the last few years progress has been made in achieving further and further *in vitro* development of implantation stages of early embryos (Hsu *et al.*, 1974), and the age at which fetuses can be carried to term outside the female reproductive tract is becoming earlier and earlier.

Forms Other Than Ova and Embryos   In theory animal germplasm could be preserved through preservation of the intact mature animal and through preservation of its somatic cells. In both cases the reduction of theory to practice remains highly speculative at best, perhaps impossible. Although a few small primitive intact metazoans and a few insects can be frozen, the size and complexity of most metazoans preclude for the foreseeable future their surviving freezing and thawing. But many animal somatic cells can be easily frozen. What remains highly speculative is whether ways can be found to derive a whole animal from some of its component somatic cells.

Plant Germplasm

By contrast with animal systems, the problem of deriving a whole plant from samples of its somatic cells seems well on the way to solution in an increasing number of species. Because attempts to freeze plant-tissue cultures were initiated only recently, we are not sure whether the task will be difficult or simple. But since it has been achieved in several plant systems, it seems reasonable to assume that the obstacles will not be insurmountable. If the assumption is correct, the combination of the preservation of tissue cultures at low temperatures with the ability to generate whole plants from the thawed material should prove a powerful approach to the preservation of plant germplasm, especially where the task is to preserve germplasm of plants that do not form seeds or that form seeds which are unable to survive long periods of storage.

8 DATA MANAGEMENT

Data on conservation of genetic resources and on the
genetic resources themselves have traditionally been the
responsibility of the individual most immediately concerned. Little concerted effort has been made to organize,
direct, collate, store, and use these data. Such an at-
titude has brought us to the point where, to a large extent,
we do not know where they are, what their quality or quantity is, what confidence we can have in them, and how they
fit into a larger scheme of knowledge about organisms.

Today some of the most valuable knowledge about genetic
resources resides in the holdings of museums, herbaria,
and similar institutions. It is sometimes difficult to
retrieve data or information from these institutions.
There is usually only one avenue to information on the
stored resources, and that is through the names of the
organisms. Should one wish to retrieve all the names of
the plants for a certain region, for example, it would be
very difficult to derive such a list from the museum or
herbarium because of the single-entry nature of the information storage system, yet inherently these institutions
possess very valuable data banks. A move to place these
data in computerized storage and retrieval systems is under
way. (The SELGEM system of the Smithsonian Institution
and the TAXIR system used at the USDA Regional Plant Introduction Station at Washington State University, Pullman,
are cases in point.) A key deterrent is lack of appreciation among biologists generally of the need for data management systems. It might be instructive to know how much
time and money is invested in data management, in light of
the casual attitudes many scientists have toward their
principal product, data.

Data management is, or should be, an integral function
associated with conservation of genetic resources. If we

do not know what we have and what we are now doing, we can hardly know whether conservation is being accomplished or whether we are merely paying lip service to some general concept. We cannot show, by reference to firm data, that any particular organism is, or is not, being lost. Aside from conspicuous plants or animals, whose existence is usually within the range of man's everyday observations, we cannot say with certainty that this or that species is on the verge of extinction. Indeed, there have been several situations in which, after an organism had been thought extinct, it proved to be alive and well in some part of the world in which we had not known it existed. Perhaps the most famous example among plants is the dawn redwood (*Metasequoia glyptostroboides*). It was long cited as a fossil in the United States and therefore thought to be extinct, but it was found during World War II in China. Shortly afterward it was cultivated in many places around the world and is no longer considered even an endangered species.

The institutions that manage information best are those with the longest record of management: zoos, botanical gardens, museums, herbaria, and live-culture collections. It is important that a critical and powerful organization committed to data management be developed.

A number of outstanding agencies, public and private, maintain systems for storing and retrieving biological literature. They include the National Agricultural Library, the National Library of Medicine, BioSciences Information Service, and Chemical Abstracts. Many private organizations that serve as information sources cannot accept responsibility for management of raw data. Their financial arrangements require that they recover costs, and there is little assurance that the market for genetic resources raw data would be dependable enough to ensure a profit. Both the National Agricultural Library and the National Library of Medicine are responsible for keeping up with and disseminating information about the published literature in their respective disciplines. Very likely isolated works in pure biology do not show up in their storage and retrieval systems. In any case, these institutions would be unable to accept unpublished data on conservation of genetic resources. Such data are merely descriptions of the actual plants or animals that are maintained or conserved.

The critical data are those describing the what, where, who, when, and why of the conserved species and genotypes. Data about the organizations and individuals carrying on

conservation work are also important. Both types of data
should be continuously updated. Standards of description,
such as those existing in chemistry and physics, are needed
and should be developed, as are networks for exchange of
data. Many institutions are without funds needed to update
their system of recording, storing, retrieving, and trans-
mitting data and materials.

Continuous research into data management for all types
of biological data is required, including not only information storage and retrieval systems, per se, but also
programs for determining diversity within species, such
multivariate analysis such as clustering methods, and
computer-aided graphics for mapping and contouring.

A first-class training program for students and research workers is needed, to ensure that they understand
procedures and the significance of accurate observation
and recording. Many new methods are available for gathering data--from remote sensing, to electron microscopy, to
protein analyzers--but it is often difficult to use these
effectively in the conservation of genetic resources.

A number of institutions have set out to automate
their data storage and retrieval systems for systematic
collections and will continue this work. At the moment
there is little or no coordination among organizations.
Computer programs are being used on several types of computers; opinions differ as to whether this or that program
is best for the purpose. We are badly in need of agreement with respect to some minimal set of descriptors for
the accessed materials. If the arguments could be directed
toward establishing the proper descriptor set, it would
help greatly. Most sophisticated computer-aided programs
require a set of data that has been structured to fit in
certain ways and is adaptable to different machines and
differently structured data. Standardization of the data
structures on magnetic tape would be very beneficial.

9 FINDINGS AND RECOMMENDATIONS

GENERAL

● The diversity of germplasms is an essential national resource.

● The diverse germplasms are currently represented in a wide variety of forms including: free-living organisms in a great diversity of natural habitats in terrestrial, marine, and freshwater ecosystems; economically important organisms, including field, vegetable, nursery, florist, fruit and nut crops; forests; drug plants; livestock; fisheries; industrial microorganisms and pathogenic bacteria; specialized collections in zoos, aquariums, and botanical gardens that preserve and study rare species; organisms used in research, including specialized genetic stocks.

● The value of these resources is being rapidly eroded by a variety of encroachments.

● Action must be taken soon to protect and maintain the remaining genetic diversity, because once lost, much of it can never be recovered.

● The National Academy of Sciences should appoint a continuing Committee on Preservation of Germplasm Resources that would provide general surveillance on germplasm maintenance, and recommend specific attention to current and potential threats to genetic diversity.

● New agencies should be established, or existing agencies charged, with the preservation of particular germplasm resources: ecosystems, individual species of recognized significance, and gene pools of special value.

● Funds should be provided to support these agencies and to train the personnel necessary for the maintenance of these essential resources.

95

NATURAL ECOSYSTEMS

● Preserve carefully chosen existing natural habitats. These habitats, developed through millions of years of evolution, contain a wealth of valuable material that we understand only dimly and that we must preserve to protect the future of mankind.

● Protect extensive areas of existing natural ecosystems. A national inventory of these ecosystems and their components must be developed. We do not even have complete inventories of species in each ecosystem, and our current incomplete understanding makes it impossible to predict results of species loss.

● Enact laws that are designed to preserve total environments intact, with less specific emphasis on the endangerment of individual species. By far the best site for conservation of endangered species is within their intact natural habitats.

● Study natural ecosystems more fully. A scheme must be devised to evaluate the uniqueness of each particular habitat and to estimate the minimum area required to support the maintenance of an adequate gene pool of each component species. This system must be designed to provide an intelligent basis for decisions on preservation and evaluation. A master plan should be developed for the maintenance of each protected area.

● Encourage provision of manpower for the study of natural ecosystems, training of ecologists, geneticists, and taxonomists.

● Establish a National Ecological Reserve Board, backed by State Boards charged with study of regional problems.

● Recognize that preservation of natural habitats is a supranational problem; that we should have special concern for preservation of tropical ecosystems.

● Give particular attention to two categories of land-based areas: endangered habitats, to be held as public trusts; threatened habitats.

● Provide special protection for freshwater natural ecosystems (springs, streams, lakes, rivers, and associated wetlands) because of the pressures upon them of agricultural, residential, and industrial use. Aquatic habitats must be intensively studied and characterized, with inventory of species. Certain critical watershed systems must be preserved as essential natural habitats.

● Expand study of marine ecosystems. Research on characterization of marine ecosystems, and inventory of the component species of each system, is essential. Study of life-histories of component species is important. Research on methods for laboratory culture and on the genetics of marine species should be encouraged. Communication between investigators studying marine ecosystems and government agencies responsible for enforcing protection of our sea and shore resources must be improved.

COLLECTIONS OF ECONOMICALLY IMPORTANT ORGANISMS

• Reemphasize concern expressed in *Genetic Vulnerability of Major Crops,* and elsewhere, for the status of our major food and fiber crops. Most of these have been developed from plants introduced from indigenous agriculture in other parts of the world. The genetic underpinning of varieties currently grown in the United States is dangerously narrow, and these varieties have proved susceptible to changes in exposure to pathogens or in environmental conditions.

• Make permanent and expand the National Plant Genetic Resources Board, and encourage cooperation with the International Board for Plant Genetic Resources, established by the Consultative Group on International Agricultural Research. National and international collaboration in the exploration, collection, conservation, documentation, and use of plant resources must be established on a firm and permanent basis.

• Support the aims and activities of the National Plant Germplasm Committee to: (a) develop a plan of repositories for clonally propagated plants; (b) provide facilities, staff and funds for USDA Inter-regional and Regional Plant Introduction Stations; (c) establish a facility for growing genetic collections of tropical plants; (d) set criteria as to the duties and responsibilities of curators of specific genetic crop collections; (e) select such curators, provide for their support, and guarantee their succession in case of retirement or death; (f) characterize major collections, with computerized data on plant genetic collections; (g) identify and rectify gaps in major collections; and (h) support research on techniques for maintaining collections.

• Establish nationally organized programs in each community or natural grouping of commodities to provide gene flow from genetic resource to cultivar. This would involve collection, maintenance, characterization and use of plant genetic resources, including information on taxonomy; cytology and cytogenetics and biochemistry; screening for useful qualities; determination of inheritance patterns; development of improved breeding stocks and commercial cultivars; and supply of seeds or other propagules to producers.

● Maintain areas of indigenous subsistence agriculture of the antecedents of major U.S. crops at their geographical sites of origin. This activity should be promoted in the immediate future, since areas of subsistence agriculture in lesser-developed countries are currently diminishing markedly, due to increased industrialization, incursion of roads, replacement by modern techniques and high-yielding strains, and increase in large-scale monoculture.

● Preserve and support valuable foreign genetic collections, such as South and Central American collections of varieties of *Zea mays*. They should be subsampled and duplicates put in the National Seed Storage Laboratory. In this effort it is very important to avoid false confidence as to permanence of present arrangements.

● Promote study and preservation of the genetic resources underlying export crops of the lesser-developed countries. This includes tea, coffee, cassava, rubber, teak, and other forest products.

● Expand research on continual maintenance of high genetic diversity in forest management, so that long-lived trees can adequately meet many environmental fluctuations during their life-histories.

● Establish large wooded areas as Forest Genetic Reserves, in which seed collection only, but not logging, would be allowed and encouraged in order to maintain ancestral tree types, and to ensure a broad genetic base for future selection. Maximum genetic diversity of forest trees should be preserved, as it is impossible to predict the future needs of commercial forestry. The possibility of allowing forest seed collection in certain wilderness areas, national forests, and national parks, should be explored.

● Develop an organized collection of crop pests and pathogens and provide identification service and advice to those working in agriculture.

● Preserve natural habitats of important drug plants, as this remains the best way of assuring future supply of currently used drug plants and potential sources of useful new drug species. Continued availability of drug plants is important to the United States, since now and in the foreseeable future some 25 percent of drug prescriptions contain one or more important plant-derived chemical constituents.

• Preserve rare breeds of livestock to provide a reservoir of unique genetic traits needed for crossbreeding
and selection of experiments to provide pertinent new
breeds in the United States. Genetic flexibility may be
especially important if feeding patterns are substantially
changed in the future to conserve grain supplies. These
stocks are in danger of extinction, and many are found
only outside of the United States. Importation, seriously
impeded at present by legal restrictions stemming from fear
of exotic disease, must be facilitated by cooperative international efforts involving both breeders and quarantine
experts, to protect future food supply. Special attention
must be given in the near future to importation of exotic
breeds of sheep and swine.

• Preserve and study marine habitats and ecosystems,
placing particular emphasis on the habitats of commercial
species and ecosystems characterized by unusually high
diversity. Inaugurate genetic studies on selected marine
species, in search of strains that can meet alterations
in ecosystems.

• Support research on genetics and aquaculture of
marine invertebrates. The commercial importance of lobsters, shrimps, clams, scallops, and oysters indicates the
desirability of developing methods for their aquaculture
and improving the understanding of the biology of these
species. This should include characterization of genetic
stocks and laboratory breeding experiments. The United
States lags far behind other developed countries in this
regard.

ECONOMICALLY IMPORTANT MICROORGANISMS

• Continue U.S. cooperation with the World Health
Organization in its efforts at the international level
to conserve germplasm resources of microorganisms pathogenic to humans.

• Organize currently diverse collections of pathogenic
microorganisms in the United States under some common aegis
to facilitate exchange of information, to provide continuity of needed collections, and to assure the general
availability of organisms in existing collections.

• Identify organizational and managerial obstacles to
conservation of our national resources of microbial germplasm in existing collections. The goal is to preserve
efficiently, safely and economically, valuable materials
that represent a major social investment. At present

there is inadequate information exchange, availability of
materials, and continuity of preservation. A single mammoth public collection is probably not the most effective
resolution of the issue.

• Establish a coordinating agency to identify major
collections, assign responsibility for their preservation
and general distribution, help recruit curators and assure their continuity, maintain quality by assembling
advisory bodies and to identify, locate, and assure support of smaller unique collections.

COLLECTIONS IN BOTANICAL GARDENS, ARBORETA, ZOOS, AND
AQUARIA

• Recognize and utilize the potentialities of botanical
gardens, arboreta, zoos, and aquaria in the conservation
of rare, threatened, and endangered species. These institutions provide supporting environments for rare species
second only to those found in their native habitats, which
often have become severely limited in extent or, at times,
completely destroyed.

• Support, under international direction, an authori-
tative worldwide inventory of rare species, their distributions and population sizes. This official list should
include rare and endangered marine and estuarine species,
and should provide the documentary basis for assigning a
species to a threatened or endangered category.

• Encourage and support activities of a central organization, such as the International Union for the Conservation of Nature, to coordinate the activities of societies
seeking to conserve particular species; to ensure that all
necessary conservation activities are undertaken; to encourage efforts to breed rare species in captivity, including exchanges between institutions; to establish
priorities, with decisions arrived at by qualified scientists, relating to species conservation efforts; and
to work for continuity in funding of these efforts.

• Encourage research on the genetics and reproductive
biology of threatened species in zoos, aquaria, botanical
gardens, and arboreta. Sperm, ova, embryos, cell strains,
and tissues from vanishing species should be saved as
future sources of information on their characteristics.
In many cases, additional research will be needed to de-
vise appropriate procedures. These efforts may well be
greatly facilitated by the establishment of research
associations between given zoos or botanical gardens and
nearby universities.

• Develop alternative approaches to facilitate necessary research and yet fulfill the intent of current laws
and regulations concerning endangered species. Research
on rare, threatened, and endangered species is severely
handicapped by the formidable administrative difficulties
of complying with these regulations.

ORGANISMS FOR RESEARCH

Genetic Stocks and Centers

• Continue maintenance of essential genetically defined
stocks of organisms, including collections of viruses and
bacteria, protozoa, algae, fungi, higher plants, insects
(particularly *Drosophila*), amphibia, birds, and experimental mammals. Each defined stock reflects considerable
effort; the organisms themselves have value commensurate
with their degree of genetic definition and with the
scientific knowledge that has resulted from their use.
Small collections, the product of one investigator's work,
are especially vulnerable beyond the founder's retirement.
Larger collections of widely used species are often organized into stock centers, so designed as to maintain a
large and defined portion of known genetic variation to
supply organisms, information, and service to many investigators.
• Assign responsibility, as curators, to appropriate
concerned geneticists, who have detailed knowledge of and
interest in the organisms in a particular collection.
Scientific boards, with members having pertinent expertise,
should be appointed to advise upon each genetic stock
collection. Maintenance of specialized collections or
centers is preferable to combination of many collections
in a single institution.
• Encourage and adequately support the stock inventory,
registry and evaluation program (Committee on Maintenance
of Genetic Stocks) of the Genetics Society of America,
to provide coordination of effort and to assure continuity,
security, and reliability. Support by the National Science
Foundation of important Genetic Stock Centers, and maintenance of germplasm resources through support from the
NIH Division of Research Resources, is very helpful, and
should be continued and expanded. Other agencies should
consider adopting the policy of direct support of genetic
stocks to assure their continuing availability, not necessarily tied to support of research that utilizes a particular stock.

• Establish an advisory group for Mammalian Genetic
Stocks, to set priorities for continuation of threatened
mutants; to establish liaison groups to facilitate effective exploitation of new mutants; to promote establishment of mutants on appropriate uniform genetic backgrounds;
and to promote cooperation among agencies (including components of the National Institutes of Health) for the
maintenance of critically needed stocks.

• Continue, expand, and diversify the very useful
organizational and informational activities of the Institute of Laboratory Animal Resources (NRC).

Organisms Obtained from the Wild

• Promote and support experimentation to develop
optimal methods for culture and reproduction of wild
species in the laboratory. Repeated harvesting from the
wild has been the chief source of such critically needed
species as echinoderms, various amphibians, various
aplysiid mollusks, and nonhuman primates. Laboratory
culture will go far to assure future supply, avoid perturbations of natural habitats, and sharpen the utility
of these research tools.

• Protect populations of nonhuman primates in the
wild and establish effective breeding colonies in the
United States. For each important species, research on
reproductive physiology and nondestructive genetic analysis
should be encouraged. Especially during the impending
period of restricted supply, while breeding colonies are
building slowly, nonhuman primates should be used only for
the most essential research, and carefully organized multiple use of scarce animals should be encouraged.

CRYOBIOLOGICAL PRESERVATION

• Encourage and support extensive research on the
freezing and resuscitation of cells. This effort must
include funds to examine the fundamental aspects of the
problems as well as for application of these fundamentals
to the development of pragmatic methods suitable to specific cell types.

• Seek better approaches for bringing the necessary
talents into juxtaposition appropriate to the strong inter-
disciplinary nature of research in cryobiology and its
applications require.

● Seek experimental confirmation of the generally held view that low-temperature storage provides an optimum method for maintaining genetic constancy of specific stocks. Theory indicates that mutation rate at liquid nitrogen temperatures will be much lower than at room temperature.

● Provide adequate support and protection for collections of frozen semen or embryos to maintain rare varieties of livestock, of frozen seeds, or other fruiting bodies of rare plant varieties. These may be an excellent way of preserving genetic diversity needed in future selection experiments.

MANAGEMENT OF DATA

● Involve biological scientists in designing data collection schemes for both the conservation process and genetic resources, using as consultants individuals with systems training.

● Develop a critical and powerful system of data management to accompany each conservation effort. Responsibility for seeing that adequate original data are collected should continue to lie with the concerned institution or governmental agency (NIH, USDA, USDI).

● Establish an overall monitoring group, with relatively long-term membership, to coordinate activities and maintain communication between different conservation efforts, and to assume general responsibility for quality and completeness of coverage.

● Provide a training program for students and research workers to ensure understanding of suitable procedures and of the significance of accurate observation and recording.

REFERENCES

Agricultural Research Policy Advisory Committee. 1973.
Recommended actions and policies for minimizing the
genetic vulnerability of our major crops. USDA and
National Association of State Universities and Land
Grant Colleges, Washington, D.C. 33 p.

Asahina, E. 1966. Freezing and frost resistance in insects, p. 451-486. *In* G. T. Meryman [Ed.] Cryobiology.
Academic Press, Inc., New York.

Ashwood-Smith, M. J. and G. Friedman. (In press)
Genetic stability in cellular systems stored in the
frozen state. *In* The Freezing of Mammalian Embryos.
Ciba Foundation Symposium, London, Jan. 1977.

Ashwood-Smith, M. J. and E. Grant. 1976. Mutation induction in bacteria by freeze-drying. Cryobiology
13:206-213.

Ashwood-Smith, M. J., J. Trevino, and C. Warby. 1972.
Effect of freezing on the molecular weight of bacte-
rial DNA. Cryobiology 9:141-143.

Austin, C. R. 1973. Embryo transfer and sensitivity to
teratogenesis. Nature 244:333-334.

Bajag, Y. P. S. 1976. Regeneration of plants from cell
suspensions frozen at -20, -70, and -196°C. Physiol.
Plant 37:263-268.

Barber, J. C. and S. L. Krugman. 1974. Preserving forest
tree germ plasm. American Forests 80:8-11, 42.

Bean, M. J. 1977. The Evolution of National Wildlife
Law. Prepared for the Council on Environmental Quality
by the Environmental Law Institute. U.S. Government
Printing Office, Washington, D.C. 485 p.

Benirschke, K. 1977. Experimental systems: advantages
and disadvantages, p. 58-75. *In* The Future of Animals,
Cells, Models, and Systems in Research, Development,
Education, and Testing. National Academy of Sciences,
Washington, D.C.

Bereskin, B. 1976. Preservation of germ plasm--an over-
view. U.S. Department of Agriculture, ARS, Agricultural
Research Center, Beltsville, Md.

Bermant, G. and D. G. Lindburg. 1975. Primate Utilization and Conservation. Wiley-Interscience, New York.

Bowman, J. C. 1974. Conservation of rare livestock breeds
in the United Kingdom. Proc. 1st Congress on Genetics
Applied to Livestock Production 2:23-29.

Bridgewater, D. C. 1972. Saving the lion marmoset. Wild
Animal Propagation Trust, Oglebay Park, Wheeling, W. Va.

Burke, M. J., L. V. Gusta, A. Quamme, C. J. Weiser, and
P. H. Li. 1976. Freezing and injury in plants. Ann.
Rev. Plant Physiol. 27:507-528.

Callaham, R. Z. 1970. Geographic variation in forest
trees, p. 43-47. O. H. Frankel and E. Bennett [Eds.]
In Genetic Resources in Plants--Their Exploration and
Conservation. IBP Handbook No. 11. Blackwell Scientific Publications, Oxford. 554 p.

Ching, T. M. and W. H. Slabaugh. 1966. X-ray diffraction
analysis of ice crystals in coniferous pollen. Cryobiology 2:321-327.

Ciba. (In press) The Freezing of Mammalian Embryos. Ciba
Foundation Symposium, London, Jan. 1977.

Committee for the Preservation of Natural Conditions.
1926. V. E. Shelford [Ed.] Naturalist's Guide to the
Americas. The Ecological Society of America. Williams
& Wilkins Co., Baltimore. 761 p.

Committee on Genetic Vulnerability of Major Crops, Agricultural Board. 1972. Genetic Vulnerability of Major
Crops. National Academy of Sciences, Washington, D.C.
307 p.

Darnell, R. M., P. C. Lemon, J. M. Neuhold, and G. Carleton.
1974. Natural areas and their role in land and water
resources preservation. Final report to the National
Science Foundation, US/IBP Program for Conservation of
Ecosystems. American Institute of Biological Sciences,
Rosslyn, Va. 286 p.

Diamond, J. M. 1976. Island biogeography and conservation: strategy and limitations. Science 193:1027-
1029.

Director General, WHO. 1969. Medical Research Programme
of the World Health Organization, 1964-1968. Report
by the Director General. WHO, Geneva. 350 p.

Dougall, D. K. and D. F. Wetherell. 1974. Storage of
wild carrot cultures in the frozen state. Cryobiology
11:410-415.

Edwards, S. R. and G. R. Pisani [Eds.] 1976. Endangered

and Threatened Amphibians and Reptiles in the United States. Society for the Study of Amphibians and Reptiles, Lawrence, Kansas. 65 p.

FAO. 1967. Report of the FAO study group on the evaluation, utilization and conservation of animal genetic resources. Food and Agriculture Organization of the UN, Rome, Italy, 21-25 November 1966 Meeting Report AN 1966/69.

FAO. 1969. Report of the second ad hoc study group on animal genetic resources. Food and Agriculture Organization of the UN, Rome, Italy, 18-22 November 1968 Meeting Report AN 1968/8.

FAO. 1971. Report of the third ad hoc study group on animal genetic resources (pig breeding). Food and Agriculture Organization of the UN, Copenhagen, Denmark, 19-24 April 1971 Meeting Report AGA 1971/3.

FAO. 1973. Report of the fourth FAO expert consultation on genetic resources (poultry breeding), Centre de Recherches de l'INRA, Nouzilly, France, 19-24 March 1973. Food and Agriculture Organization of the UN, Rome, Italy, Meeting Report AGA 1973/1.

Farnsworth, N. R. and R. W. Morris. 1976. Higher plants; the sleeping giant of drug development. Am. J. Pharm. 148:46-52 (cited from p. 54, National Prescription Audit, published by R. A. Gosselin and Co., 1976).

Farrant, J., C. A. Walter, and S. C. Knight. 1977a. Cryopreservation and selection of cells, p. 61-78. D. Simatos, D. M. Strong, and J. M. Turc [Eds.] In Cryoimmunology. L'Institut National de la Santé et de la Recherche Médicale, Paris. 423 p.

Farrant, J., C. A. Walter, H. Lee, and L. E. McGann. 1977b. Use of two-step cooling procedures to examine factors influencing cell survival following freezing and thawing. Cryobiology 14:273-286.

Fish and Wildlife Service. 1977. Endangered Species Technical Bulletin 2(2) February 1977. U.S. Government Printing Office, Washington, D.C. 4 p.

Fitter, R. 1977. (Editorial Notes) Oryx 13:440.

Franklin, J. F. 1977. The biosphere reserve program in the United States. Science 195:262-267.

Fry, R. M. 1966. Freezing and drying of bacteria, p. 665-696. H. T. Meryman [Ed.] In Cryobiology. Academic Press, New York.

Genoway, H. H. and J. R. Choate. 1976. Federal regulations pertaining to collection, import, export, and transport of scientific specimens of mammals. J. Mammal. 57(2) Suppl.:1-9.

108

Gooch, J. L. 1975. Some current problems in marine genetics, p. 85-107. J. D. Costlow [Ed.] *In* The Ecology
of Fouling Communities. U.S. Office of Naval Research.
Goodwin, R. E. and W. A. Niering. 1975. Inland Wetlands
of the United States: Evaluated as Potential Registered
Natural Landmarks. National Park Service. U.S. Govern-
ment Printing Office, Washington, D.C. 550 p.
Graham, E. F. 1973. The state of the art in cryobiology,
p. 182-189. *In* Applications of Cryogenic Technology,
Vol. 5. Scholium International, Inc., Flushing, N.Y.
Graham, E. F. 1976. Studies of supercooling, freezing,
and osmotic pressure on motility and fertility of
turkey spermatozoa. *In* Annual Research Report,
Minnesota Turkey Research and Development Board.
Haight, M., J. Frim, J. Pasternak, and H. Frey. 1975.
Freeze-thaw survival of the free-living nematode
Caenorhabditis briggsae. Cryobiology 12:497-505.
Harlan, J. R. 1975. Our vanishing genetic resources.
Science 188:618-622.
Hellawell, J. M. 1977. Changes in natural and managed
ecosystems: detection, measurement and assessment.
Proc. Roy. Soc. Lond. B 197:31-57.
Holm-Hansen, O. 1963. Viability of blue-green and green
algae after freezing. Physiol. Plant 16:530-540.
Horton, H. F. and A. G. Ott. 1976. Cryopreservation of
fish spermatozoa and ova. J. Fisheries Research Board
of Canada 33:995-1000.
Hsu, Y.-C., J. Baskar, L. C. Stevens, and J. E. Rash. 1974.
Development *in vitro* of mouse embryos from the two-cell
stage to the early somite stage. J. Embryol. Exp.
Morph. 31:235-245.
Hubbs, K. R. and J. Bleby. 1976. Laboratory non-human
primates for biomedical research in the United Kingdom.
A report to the medical research council on the exist-
ing provision and future needs. Medical Research
Council, Surrey, England.
Institute of Laboratory Animal Resources. 1975a. Research
in Zoos and Aquariums. A Symposium. National Academy
of Sciences, Washington, D.C. 215 p.
Institute of Laboratory Animal Resources. 1975b. Nonhuman
Primates: Usage and Availability for Biomedical Programs. National Academy of Sciences, Washington, D.C.
122 p.
Institute of Laboratory Animal Resources. 1976. Federal
regulations pertaining to the collection, import, export,
and transport of scientific specimens and mammals. ILAR
News 20:15-19.

109

International Union for the Conservation of Nature and Natural Resources (IUCN). 1972. Red Data Books. Vol. I, Mammalia. Morges, Switzerland.

Janzen, D. H. 1973. Tropical agroecosystems. Science 182:1212-1219.

Jenkins, R. E. (In press) Classification and inventory for the perpetuation of ecological diversity. Report presented to a national symposium on the classification, inventory, and analysis of fish and wildlife habitat sponsored by the Department of the Interior, U.S. Fish and Wildlife Service, January 24, 1977.

Jewell, P. A. 1971. The case for the preservation of rare breeds of domestic livestock. Veterinary Record 85:524-527.

Karow, A. M., Jr., G. J. M. Abouma, and A. L. Humphries. 1974. Organ Preservation for Transplantation. Little Brown, Boston.

Keilen, D. 1959. The problem of anabiosis or latent life: History and current concepts. Proc. Roy. Soc. London B 150:149-191.

Kifer, J. F. 1975. NOAA's marine sanctuary program. Coastal Zone Management Journal 2:177-188.

Kinne, O. and H. P. Bulnheim [Eds.] 1970. Cultivation of marine organisms and its importance for marine biology. International Helgoland Symposium, 1969. Helgoländer Wissenschaftliche Meeresuntersuchungen 20:1-721.

Koehler, J. K. and L. K. Johnson. 1969. Food supply as a factor in the survival of frozen and thawed rotifers. Cryobiology 5:375-378.

Küchler, A. W. 1964. Potential natural vegetation of the conterminous United States. American Geographical Society Special Publication No. 36.

Lauvergne, J. J. 1975. Conservation of animal genetic resources (pilot study). FAO Project No. 0604-73/002, Rome, Italy.

Leef, J. L. and F. H. Gaertner. 1975. A cryobiological method for isolating fungal mutants. Cryobiology 12:584.

Leibo, S. P. (In press) Fundamental cryobiology of mouse ova and embryos. In The Freezing of Mouse Embryos. Ciba Foundation Symposium, 1977.

Leibo, S. P. and P. Mazur. 1970. Mechanisms of freezing damage in bacteriophage T4, p. 235-246. G. Wolstenholme and M. O'Connor [Eds.] In The Frozen Cell. Ciba Foundation Symposium, Churchill, London.

Levitt, J. 1956. Hardiness of Plants. Academic Press, New York. 278 p.

110

Levitt, J. 1964. Cryobiology as viewed by the botanist. Cryobiology 1:11-17.

Lowdermilk, W. C. 1953. Conquest of the land through seven thousand years. U.S. Department of Agriculture Information Bull. No. 99. U.S. Government Printing Office, Washington, D.C. 30 p.

Lyon, M. F., D. G. Whittingham, and P. Glenister. (In press) Long-term storage of frozen mouse embryos under increased background irradiation. *In* The Freezing of Mammalian Embryos. Ciba Foundation Symposium, London, Jan. 1977.

Maini, J. S. 1973. Conservation of forest tree gene resources in Canada: An ecological perspective, p. 43-50. D. P. Fowler and C. W. Yeatman [Eds.] *In* Proceedings of the Thirteenth Meeting of the Committee on Forest Tree Breeding in Canada, Part 2. Canadian Forestry Service, Ottawa. 86 p.

Maini, J. S., C. W. Yeatman, and A. H. Teich. 1975. In situ and ex situ conservation of gene resources of *Pinus banksiana* and *Picea glauca*. FO. Misc-75-8. p. 27-40. *In* Report on a pilot study on the methodology of conservation of forest genetic resources. Food and Agriculture Organization of the UN, Rome, Italy.

Martin, A. C., N. Hotchkiss, F. M. Uhler, and W. S. Bourn. 1953. Classification of wetlands of the United States by the Waterlands Classification Committee of the Fish and Wildlife Service. U.S. Department of the Interior, Fish and Wildlife Service, Washington, D.C. 14 p.

Martin, P. S. 1966. Africa and Pleistocene overkill. Nature 212:339-342.

Martin, R. D. 1975. Breeding endangered species in captivity. Academic Press, London. 420 p.

Martin, S. M., and C. Quadling. 1970. World directory of culture collections, p. 133-139. H. Iizuka and T. Hasegawa [Eds.] *In* Proceedings of the International Conference on Culture Collections, Tokyo, October 7-11, 1968. University Park Press, Baltimore. 625 p.

Martin, S. M., and V. B. D. Skerman [Eds.] World directory of collections of cultures of microorganisms. World Federation for Culture Collections of the International Association of Microbiological Societies. Wiley Interscience, John Wiley and Sons, Inc., New York. 560 p.

Mason, I. L. 1974. The conservation of animal genetic resources: Introduction to round table. Proceedings First World Congress on Genetics Applied to Livestock Production 2:13-21.

111

Maurer, R. R., H. Bank, and R. E. Staples. 1977. Pre-
and postnatal development of mouse embryos after storage
for different periods at cryogenic temperatures.
Biology Reproduction 16:139-146.

May, R. M. 1973. Stability and complexity in model
ecosystems. Princeton University Press, Princeton,
N.J. 235 p.

Mayr, E., R. D. Alexander, W. F. Blair, P. Illg, B.
Schaeffer and W. C. Steere. 1974. The diversity of
life, p. 20-49. W. H. Johnson and W. C. Steere [Eds.]
In The Environmental Challenge. Holt, Rinehart and
Winston, Inc., New York.

Mazur, P. 1966. Physical and chemical basis of injury
in single-celled micro-organisms subjected to freezing
and thawing, p. 213-215. H. T. Meryman [Ed.] In
Cryobiology. Academic Press, Inc., New York.

Mazur, P. 1968. Survival of fungi after freezing and
desiccation, p. 325-394. G. Ainsworth and A. S.
Sussman [Eds.] In The Fungi, Vol. III, Chapter 14.
Academic Press, New York.

Mazur, P. 1969. Freezing injury in plants, p. 419-448.
Annual Review of Plant Physiology, Vol. 20. Annual
Reviews, Inc., Palo Alto, Calif.

Mazur, P. 1970. Cryobiology: The freezing of biological
systems. Science 168:939-949.

Mazur, P. 1976. Freezing and low temperature storage of
living cells, p. 1-12. Otto Mühlbock [Ed.] In
Proceedings of the Workshop on Basic Aspects of Freeze
Preservation of Mouse Strains, Jackson Laboratory, Bar
Harbor, Maine, September 16-18, 1974. Gustav Fischer
Verlag, Stuttgart.

Mazur, P. 1977. The role of intracellular freezing in
the death of cells cooled at supraoptimal rates. Cryobiology 14:287-272.

Mazur, P., J. A. Kemp, and R. H. Miller. 1976. Survival
of fetal rat pancreases frozen to -78 and -196°C.
Proc. Nat. Acad. Sci. 73:4105-4109.

Meryman, H. T. [Ed.] 1966a. Cryobiology. Academic Press,
Inc., New York.

Meryman, H. T. 1966b. Freeze-drying in cryobiology, p.
609-663. H. T. Meryman [Ed.] In Cryobiology,
Academic Press, Inc., New York.

Meryman, H. T., R. J. Williams, and M. St. J. Douglas.
1977. Freezing injury from "solution effects" and its
prevention by natural or artificial cryoprotection.
Cryobiology 14:287-302.

Miller, R. R. 1967. Status of native fishes of the
Death Valley System in California and Nevada. Completion

112

Report of Resource Studies Problem undertaken for the
National Park System. National Park System DEVA-67.
Moyseenko, H. P., J. L. Woodall, and S. A. Woodall. (In
press) A balanced ecogeographical information system:
a vehicle for data collection, systematization and dis-
semination. Report presented to a national Symposium
on the Classification, Inventory and Analysis of Fish
and Wildlife Habitat sponsored by the Department of
the Interior, U.S. Fish and Wildlife Service, January
24-27, 1977.
Mühlbock, O. 1976. Proceedings of the Workshop on Basic
Aspects of Freeze Preservation of Mouse Strains.
Gustav Fischer Verlag, Stuttgart. 133 p.
Murphy, L. S. and R. R. L. Guillard. 1976. Biochemical
taxonomy of marine phytoplankton by electrophoresis
of enzymes. I. The centric diatoms *Thalassiosira
pseudonana* and *T. fluviatilis*. J. Phycol. 12:9-13.
Myers, N. 1977. Garden of Eden to weed patch. *In*
Natural Resources Defence Council Newsletter, Vol. 6,
#1, Jan./Feb.
Nace, G. W. 1976. Standards for laboratory amphibians.
K. V. Fite [Ed.] The Amphibian Visual System--A Multi-
disciplinary Approach. Academic Press, New York.
Nag, K. K. and H. E. Street. 1975. Freeze-preservation
of cultured-plant cells. II. The freezing and thaw-
ing phases. Physiol. Plant 34:216-265.
Nanney, D. L. and J. W. McCoy. 1976. Characterization
of the species of the *Tetrahymena pyriformis* complex.
Trans. Amer. Microscopical Soc. 95:664-682.
The Nature Conservancy. 1975. The preservation of
natural diversity: A survey and recommendations. 212
p. & 94 p. appendices.
Nyberg, D. 1973. Breeding systems and resistance to
environmental stress in ciliates. Evolution 28:367-380.
Odum, H. T., B. J. Copeland and E. A. McMahan [Eds.] 1974.
Coastal Ecological Systems of the United States, Vol. I.
The Conservation Foundation, Washington, D.C. 533 p.
Olney, P. J. S. 1976. International Zoo Yearbook. Vol.
16. Zoological Society, London. International Publi-
cations Service, New York.
Olney, P. J. S. 1977. Breeding Endangered Species in
Captivity. International Zoo Yearbook. Vol. 17.
Zoological Society, London. International Publications
Service, New York. 122 p.
Pegg, D. G. 1970. Banking of cells, tissues, and organs
at low temperatures, p. 153-180. A. U. Smith [Ed.]
In Current Trends in Cryobiology. Plenum Press, New York.

Perlman, D. 1977. Fermentation industries--Quo vadis? Chem. Tech. 7:434-443.

Prance, G. T. and T. S. Elias. 1977. Extinction is Forever. New York Botanical Garden, New York. 437 p.

Provasoli, L. 1976. Keynote Address: Nutritional aspects of crustacean aquaculture, p. 13-21. K. S. Price, Jr., W. N. Shaw, and K. S. Danberg [Eds.]. Proceedings of the First International Conference on Aquaculture Nutrition, October 14-15, 1975, Lewes/Rehoboth, Delaware.

Rajotte, R. V., J. B. Dossetor, and W. A. G. Voss. 1976. Frozen fetal and neonatal mouse cardiac reimplants for assessing cryoprotective agents. Cryobiology 13:609-615.

Ray, G. C. 1975. A preliminary classification of coastal and marine environments. IUCN Paper No. 14. International Union for the Conservation of Nature, Morges, Switzerland. 26 p.

Rendel, J. 1975. The utilization and conservation of the world's animal genetic resources. Agriculture and Environment 2:101-119.

Ricklefs, R. E. 1973. Ecology. Chiron Press, Newton, Mass. 534 p.

Ridgway, S. and K. Benirschke. (In press) Dolphin breeding workshop (Conference held in 1975).

Roberts, E. H. 1975. Storage of wheat seed for genetic conservation. In International Board for Plant Genetic Resources Symposium on Wheat Genetic Resources, Leningrad, July 14-22, 1975.

Roberts, E. H. and R. H. Ellis. 1977. Prediction of seed longevity at subzero temperatures and genetic resources conservation. Nature 268:431-433.

Ross, J. F. 1974. Estaurine sanctuaries--the Oregon experience. Coastal Zone Management Journal 1:433-446.

Russell, W. L. 1963. The effect of radiation dose rate and fractionation on mutation in mice, p. 205-217. F. H. Sobels [Ed.] In Repair from Genetic Radiation--Damage and Differential Radiosensitivity in Germ Cells. Macmillan, New York and Pergamon, London. 454 p.

Seager, S. W. J., C. C. Platz, and W. S. Fletcher. 1975. Conception rates and related data using frozen dog semen. J. Reprod. Fertil. 45:189.

Seibert, M. and P. J. Wetherbee. 1977. Increased survival and differentiation of frozen herbaceous plant organ cultures through cold treatment. Plant Physiol. 59:1043-1046.

Sears, P. B. 1935. Deserts on the March. Oklahoma Press, Norman, Okla. 231 p.

114

Selander, R. K. 1976. Genic variation in natural populations, p. 21-45. F. J. Ayala [Ed.] Molecular Evolution.
Sinauer Associates, Inc., Sunderland, Mass.

Shapley, D. 1977. Antarctic problems: Tiny krill to
usher in new resource era. Science 196:503-505.

Shaw, S. P. and C. G. Fredine. 1956. Wetlands of the
United States. U.S. Department of the Interior, Fish
and Wildlife Service, Circular 39. 67 p.

Shikama, K. 1965. Effect of freezing and thawing on the
stability of double helix of DNA. Nature 207:529-530.

Short, R. V. 1976. The introduction of new species of
animals for the purpose of domestication. Londzucker-
man [Ed.] In The Zoological Society of London 1826-
1976 and Beyond. Zoological Society of London,
Symposium 40.

Simberloff, D. S. and L. G. Abele. 1976. Island biogeography theory and conservation practice. Science
191:285-286.

Simmons, E. G. 1970. Commentary on Martin and Quadling's
paper. H. Iizuka and T. Hasegawa [Eds.] In Proceedings
of the International Conference on Culture Collections,
Tokyo, October 7-11, 1978. University Park Press,
Baltimore. 625 p.

Simon, E. M. 1971. *Paramecium aurelia*: Recovery from
-196°C. Cryobiology 8:361-365.

Smith, A. U. and C. Polge. 1950. Survival of spermatozoa
at low temperatures. Nature 166:668.

Smith, W. L. and M. H. Chanley [Eds.] 1975. Culture of
Marine Invertebrate Animals. Plenum Press, New York.
338 p.

Sonneborn, T. M. 1957. Breeding systems, reproductive
methods and species problems in protozoa, p. 155-324.
E. Mayr [Ed.] In The Species Problem. American Association for the Advancement of Science, Washington, D.C.

Sonneborn, T. M. 1975. The *Paramecium aruelia* complex of
14 sibling species. Transactions of American Microscopical Society 94:155-178.

Steeves, T. A. and I. M. Sussex. 1972. The cellular
basis of organization. In Patterns of Plant Development, Prentice Hall, Inc., Englewood Cliffs, N.J.
302 p.

Stephan, C. E. 1975. Methods for acute toxicity tests
with fish, macroinvertebrates and amphibians. U.S.
Environmental Protection Agency, Corvallis, Oregon (EPA -
660/3-75-009). 61 p.

Steponkus, P. L., M. P. Garber, S. P. Myers, and R. D.
Lineberger. 1977. Effect of cold acclimation and

115

freezing on structure and function of chloroplast
thylakoids. Cryobiology 14:303-321.
Terborgh, J. 1976. Island biogeography and conservation:
strategy and limitations. Science 193:1029-1030.
Towill, L. E. and P. Mazur. 1976. Osmotic shrinkage as
a factor in freezing injury in plant tissue culture.
Plant Physiol. 57:290-296.
Tracey, M. L. and K. B. Nelson. 1975. Allozymic varia-
tion in the American lobster, *Homarus americanus*.
Genetics 80 (Suppl.):81.
Weber, W. A. and B. C. Johnston. 1976. Natural History
Inventory of Colorado. I. Vascular Plants, Lichens,
and Bryophytes. University of Colorado Museum, Boulder,
Colorado. 205 p.
Weiser, C. J. 1970. Cold resistance and injury in wood
plants. Science 169:1269-1278.
Wharton, C. H. 1970. The Southern River Swamp--A Multiple
Environment. Bureau of Business and Economic Research,
Georgia State University School of Business Administra-
tion, Atlanta, Ga. 48 p.
Whitcomb, R. F., J. F. Lynch, P. A. Opler and C. S.
Robbins. 1976. Island biogeography and conservation:
strategy and limitations. Science 193:1030-1032.
Whittingham, D. G., S. P. Leibo, and P. Mazur. 1972.
Survival of mouse embryos frozen to -196° and -269°C.
Science 178:411-414.
Wilkes, G. 1977. The world's crop plant germplasm--an
endangered resource. Bull. Atom. Sci. 33:8-16.
Yeatman, C. W. 1972. Gene pool conservation for applied
breeding and seed production. B-8(V) 1-6. Proceed-
ings of the Joint Symposia for the Advancement of
Forest Tree Breeding of Genetics Subject Group, the
International Union of Forest Research Organizations.
The Government Forest Experiment Station of Japan,
Tokyo.
Zisweiler, V. 1967. Extinct and vanishing animals.
Heidelberg Science Library, Vol. 2. Springer-Verlag,
New York. 133 p.

BIBLIOGRAPHY

ARS. 1977. The National Plant Germplasm System. Program Aid 1188. Agricultural Research Service, U.S. Department of Agriculture.

Brown, A. W. A., T. C. Byerly, M. Gibbs, and A. San Pietro [Eds.] 1975. Crop Productivity--Research Imperatives. Michigan Agricultural Experiment Station, East Lansing, Michigan, and Charles F. Kettering Foundation, Yellow Springs, Ohio. 399 p.

Bye, Robert. 1976. Ethnoecology of the Tarahumara of Chihuahua, Mexico. Ph.D. Thesis, Department of Biology, Harvard University.

Ehrenfeld, D. W. 1976. The conservation of non-resources. American Scientist 64:648-656.

Frankel, O. H. and E. Bennett. 1970. Genetic Resources in Plants--Their Exploration and Conservation. IBP Handbook No. 11. F. A. Davis, Philadelphia. 554 p.

Frankel, O. H. and J. G. Hawkes. 1975. Crop Genetic Resources for Today and Tomorrow. IBP #2. Cambridge University Press, Cambridge, UK. 492 p.

Harcombe, P. A. and P. L. Marks. 1976. Species preservation. Science 194:383.

Hwang, S. W., E. E. Davis, and M. T. Alexander. 1964. Freezing and viability of *Tetrahymena pyriformis* in dimethyl sulfoxide. Science 144:64-65.

Jenkins, R. E. 1972. A national natural areas inventory. The Nature Conservancy News 22:16-18.

Leyhausen, P. 1977. Breeding endangered species. Oryx 13:427-428.

Murtfeldt, L. 1977. AAZPA/ISIS activity accelerates. AAZPA (American Association of Zoological Parks and Aquariums) Newsletter vol. XVIII-3, p. 10. ISIS (International Species Inventory System).

118

Nanney, D. L. 1974. Aging and long-term temporal regulation in ciliated protozoa. Mechanisms of Aging and Development 5:503-512.

Ray, G. C. 1975. Critical Marine Habitats. IUCN occasional paper. International Union for the Conservation of Nature, Morges, Switzerland. 59 p.

Roberts, E. H. 1975. Problems of long-term storage of seed and pollen for genetic resources conservation, p. 269-296. O. H. Frankel and J. G. Hawkes [Eds.] Crop Genetic Resources for Today and Tomorrow. Cambridge University Press, Cambridge. 492 p.

Simon, E. M. 1972. Freezing and storage in liquid nitrogen of axenically and monoxenically cultivated *Tetrahymena pyriformis*. Cryobiology 9:75-81.

Simon, E. M. and S. W. Hwang. 1967. *Tetrahymena*: Effect of freezing and subsequent thawing on breeding performance. Science 155:694-696.

Simon, E. M. and M. V. Schneller. 1973. The preservation of free-living protozoa at low temperatures. Cryobiology 10:421-426.

Stebbins, G. L. 1972. Ecological distribution of centers of major adaptive radiation in Angiosperms. D. Valentine [Ed.] *In* Taxonomy, Phytogeography and Evolution. Academic Press, New York.

Wake, D. B., R. G. Zweifel, H. C. Dessauer, G. W. Nace, E. R. Pianka, G. B. Rabb, R. Ruibal, J. W. Wright, and G. R. Zug. 1975. Report of the Committee on Resources in Herpetology. Copeia 2:391-404.